Manual of Water Supply Practices

M22

# Sizing Water Service Lines and Meters

## Fourth Edition

AF836300

American Water Works Association

Manual of Water Supply Practices—M22, Fourth Edition

# Sizing Water Service Lines and Meters



Senior Editorial Manager – Book Products: Melissa Valentine
Senior Manuals Specialist: Molly Beach
Manager – Publishing Operations: Gillian Wink
Technical Editor: Dianne Beirne
Specialist 1 – Projects: Peggy Tyler
Cover Composition/Technical Illustrations: Michael Labruyere
Production: Innodata
Cover Image: Aleksandr_Kuzmin/Shutterstock.com

**Library of Congress Cataloging-in-Publication Data**

Names: Mayer, Peter W., editor. | American Water Works Association, editor.
Title: M22 - sizing water service lines and meters / [edited] by Peter W. Mayer.
Other titles: Sizing water service lines and meters
Description: Fourth edition. | Denver, CO : American Water Works Association, [2024] | "Prepared by a subcommittee of AWWA Distribution and Plant Operations Division"--preface. | Includes bibliographical references and index.
Identifiers: LCCN 2023015356 (print) | LCCN 2023015357 (ebook) | ISBN 9781647171247 (paperback) | ISBN 9781613006559 (adobe pdf)
Subjects: LCSH: Water consumption--Measurement. | Water-pipes. | Water-meters. | Water--Distribution--Planning.
Classification: LCC TD499 .S59 2024 (print) | LCC TD499 (ebook) | DDC 628.1/440287--dc23/eng/20230427
LC record available at https://lccn.loc.gov/2023015356
LC ebook record available at https://lccn.loc.gov/2023015357

Printed in the United States of America

ISBN: 978-1-64717-124-7

ISBN, electronic: 978-1-61300-655-9
https://doi.org/10.12999/AWWA.M22ed4



American Water Works Association

American Water Works Association
6666 West Quincy Avenue
Denver, CO 80235-3098
**awwa.org**

# Contents

# Figures

# Tables

# Preface

The American Water Works Association published the first edition of M22, *Sizing Water Service Lines and Meters*, in 1975. The manual was the first effort to provide guidance to the water industry on sizing water meters and service lines to meet the objectives of water utilities and their customers. The AWWA Distribution and Plant Operations Division recognized that the manual needed an update to incorporate changes in water demands that had occurred over the past 10 years and to offer a method for sizing dedicated irrigation meters that aligns with practices used by the Irrigation Association. In 1995, a manual revision subcommittee was established to review the existing manual and begin the process of developing an updated version and incorporating the impacts of existing technology into this sizing process. The efforts of that subcommittee resulted in the second edition (2004) of the meter and service-line sizing manual. In 2010, the Customer Metering Practices Committee formed a subcommittee to begin work on the third edition (2014) of M22. The subcommittee regrouped in 2020 to begin work on the fourth edition and met regularly over a three-year period to complete this most recent update.

This manual provides guidance on sizing water meters and service lines to meet the objectives of water utilities and their customers. The information in this manual can be used to estimate customer demand and maximum expected flow at a site, and this can be used to appropriately size a new service line and meter. This manual includes a useful field method called *demand profiling* that can be used to evaluate actual customer use patterns and help optimize meter size selection.

The data presented in the manual were obtained using a variety of methods including field measurements, utility surveys, technical publications, and hydraulic design calculations. This information has been condensed into a simplified format to assist readers in addressing most common service conditions. However, water and building systems are unique, and there may be complex meter and service line sizing situations that are beyond the scope of this manual.

This page intentionally blank.

# Acknowledgments

The AWWA Technical & Educational Council, the Distribution and Plant Operations Division, and the Customer Metering Practices Committee gratefully acknowledge the contributions made by those volunteers who drafted, edited, and provided the significant and critical commentary essential to updating M22. The Steering Committee members dedicated many hours in the final stages of preparation of this edition to ensure the overall technical quality, consistency, and accuracy of the manual.

**Steering Committee Members**
*Peter W. Mayer,* Chair, Water Demand Management, Boulder, Colo.
*Steven G. Buchberger,* University of Cincinnati, Cincinnati, Ohio
*Chris Douglas,* Hazen and Sawyer, Cincinnati, Ohio
*Neil Kaufman,* Truckee Donner Public Utility District, Truckee, Calif.
*Juneseok Lee,* Manhattan College, New York, N.Y.
*Gary B. Trachtman,* Arcadis, Birmingham, Ala.
*Tom Walski,* Bentley Systems Inc., Nanticoke, Pa.

**Contributors to the 4th edition**
*Peter W. Mayer,* Chair, Water Demand Management, Boulder, Colo.
*Roland Asp,* National Fire Sprinkler Association Inc., Linthicum, Md.
*Bradford P. Brainard,* Master Meter Inc., Hinesburg, Vt.
*Daniel P. Cole,* IAPMO, Mokena, Ill.
*John Collett,* Hazen and Sawyer, Greensboro, N.C.
*Steve Davis,* Arcadis, Phoenix, Ariz.
*Chris Douglas,* Hazen and Sawyer, Cincinnati, Ohio
*Jaclyn Gorman,* Denver Water, Denver, Colo.
*David Hale,* American Water Works Association, Denver, Colo.
*Craig C. Hannah,* Johnson Controls Inc., Lubbock, Tex.
*David M. Hughes,* Modernizing Distribution, Parker, Colo.
*Margaret Hunter,* American Water, Camden, N.J.
*Kathlie S. Jeng-Bulloch,* Engineering Department Director, Huntsville, Tex.
*Neil Kaufman,* Truckee Donner Public Utility District, Truckee, Calif.
*George A. Kunkel Jr.,* Kunkel Water Efficiency Consulting, Hershey, Pa.
*Frank S. Kurtz,* American Water Works Association, Denver, Colo.
*Juneseok Lee,* Manhattan College, New York, N.Y.
*Jennifer McGregor,* East Bay Municipal Utility District, Oakland, Calif.
*John F. Sliwa,* American Water, Camden, N.J.
*James Smith,* Sensus, a Xylem brand, Morrisville, N.C.
*Gary B. Trachtman,* Arcadis, Birmingham, Ala.
*Tom Walski,* Bentley Systems Inc., Nanticoke, Pa.

This page intentionally blank.

Chapter 1

# Introduction

## OVERVIEW

The physical interface between water providers and their individual customers begins and ends at the water meter, service connection, and service line. Ownership and responsibility for the meter and service line vary by utility and jurisdiction, but this connection is often understood as the place where utility-maintained infrastructure ends and customer-maintained infrastructure begins. This is where customer usage is measured and where utility revenue is determined.

Service connections and water meters are sized first and foremost to provide peak water demand at adequate pressure under all anticipated customer demand requirements, which may include fire protection. The size of the service connection and meter is important for other reasons as well. Connection fees are paid whenever a new service is established or when a customer requires additional capacity at an existing service. Volume-based utility revenue is determined from metered consumption, and meter accuracy is affected by flow rates and usage. Properly sizing water meters ensures revenue can be collected for all water that is delivered to end users.

This is the fourth edition of AWWA Manual M22, *Sizing Water Service Lines and Meters*. AWWA aims to produce manual updates every 10 years to keep current with industry changes. The first edition was printed in 1975, the second edition in 2004, and the third edition in 2014. This fourth edition includes a new approach for sizing meters for single-family and many multifamily residential buildings and includes results from recent research on measured peak demands. This fourth edition includes improved guidance on the sizing of water service lines and new approaches and data for the sizing of water meters.

## M22 TARGET AUDIENCE

This manual is intended for use by:

- water utility managers,

- engineers,
- planners,
- technicians and field operations personnel,
- consultants involved with designing and constructing projects requiring water service,
- academicians,
- meter manufacturers, and
- property developers.

## UPDATES

When the first edition (1975) of M22 was written, the primary guideline for projecting water demands was the fixture value method, which conservatively projected water demands based on then-existing fixtures and appliances. This method emphasized identifying the worst-case instantaneous peak demand that could occur for a given water customer by assuming all fixtures were drawing water simultaneously.

The second edition (2004) of M22 included methods for developing locally specific demand profiles to enable meters and service lines to be sized based on current local conditions that reflect changes in water use patterns that have occurred. These revisions were necessary to reflect changes in how much water fixtures used, irrigation demands, and the resulting implications on meter and service line sizing.

The third edition (2014) expanded on the second edition's approach and called for new research. One result was the 2020 Water Research Foundation (WRF) report, "Assessing Water Demand Patterns to Improve Sizing of Water Meters and Service Lines" (project 4689), which found evidence of meter oversizing (Mayer et al. 2020).

### Summary of Fourth Edition Content

This fourth edition of M22 provides a new approach for sizing residential water meters and adjusts the sizing curves for nonresidential meters. The fourth edition incorporates findings and data from the 2020 WRF project 4689 study; the 2016 "Residential End Uses of Water, Version 2" study (WRF project 4309B; DeOreo et al. 2016); and other studies including similar high-resolution demand profiling approaches. The "Residential End Uses of Water, Version 2" study provides an updated and expanded assessment of water use in single-family households across North America and presents detailed information and data about how water use has changed since WRF's oft-cited 1999 "Residential End Uses of Water" was published (Mayer et al. 1999). These results and data as well as other entities advancing the knowledge of domestic water usage were used to develop a new methodology and calculator for sizing water meters for single-family and small to mid-sized multifamily buildings. Every chapter in the manual has been updated to include the latest information and guidance, and Chapter 4 has been significantly expanded from earlier editions. A summary of each chapter follows.

**Chapter 2: Consumer water demands, trends, and considerations.** The manual is structured to establish predicted peak water demand as the fundamental factor to consider when sizing water service lines and meters. Currently, more is known and understood about general water demand patterns and peak demands because of the increasing focus on demand management in the water industry. Many water demand studies have been conducted in recent years that indicate distinctly different water demand patterns between and among various user classes, land uses, and geographic regions. For example, peak demands can vary greatly depending on the regional location (e.g., semi-arid West versus the more temperate Northeast) and seasonal variation of a particular location. Some water service areas contain several microclimates, which may impact sizing decisions.

This chapter reviews demand changes that have occurred and presents current data on water demands and fixture specifications.

**Chapter 3: Demand profiling for optimal meter sizing.** More equitable billing is achieved by right-sizing a higher percentage of meters so that water charges more accurately reflect water used. Current technology is helping water professionals to better understand the water demand conditions that are embedded in sizing decisions. Chapter 3 reviews equipment and technology that are available to help water utilities and their constituencies better understand typical peak demand patterns and how water is used in their systems. The chapter also reviews how demand measurement equipment measures real-time peak demands and outlines general guidelines for using this equipment for meter sizing purposes. Numerous utilities actively collect and use demand profile information to improve sizing decisions. Many water utilities have used demand profile information to downsize meters, reduce nonrevenue water, and bill for service more equitably in their water system. This chapter also includes examples of how utilities are using demand profiling in their operations.

**Chapter 4: Estimating peak water demands for service lines and meters.** To properly size water taps, meters, and service lines, the peak demands must be estimated for any specific tap. Chapter 4 presents significantly new and revised approaches for estimating peak demands in residential and nonresidential buildings. For nonresidential buildings equipped with flush-valve style toilets, revised demand curves based on recently completed research are presented. For residential and small buildings with tank-type toilets, a new and freely available water demand calculator based on empirical data collected through recent end-use studies is presented.

**Chapter 5: Service-line sizing.** There is a distinction between service-line sizing decisions and meter-sizing decisions. Service lines must be able to meet the instantaneous peak demand of a particular customer and any possible successor customer, preferably without impacting the level of water service to the customer. Meter sizing must primarily consider the accurate measurement for all expected flow ranges for the current customer to ensure accurate billing of consumption. The difference is that the service line should be more conservatively sized than the meter in general (as local conditions warrant).

This chapter provides the framework for local water utilities to ensure service lines are sized to meet peak demands while considering all relevant factors, including system pressure, line size, material, head loss, elevation changes, and type of use. Considerations for fire flow and fire sprinkler configurations are presented. Sample calculations are included as a guideline to assist the users in their own calculations.

**Chapter 6: Meter sizing and selection.** This chapter presents guidelines on meter selection based on actual or likely demands. The chapter includes system components review, meter selection, and metering configurations for fire services.

## CONSIDERATIONS WHEN SIZING WATER METERS AND SERVICE LINES

The following changes highlight the need to continually improve and update sizing methods.

### Changes in Technology

Existing water meters are technological products, some with advanced reading capabilities. Meter technology for both measuring volume and transmitting measurements is changing, and this manual includes important updates such as the inclusion of electronic meters.

Technology changes are also reducing average and peak demands. This fourth edition of M22 has also taken these factors into account.

## Demand Pattern Changes

Demand management has been institutionalized through national energy and water efficiency standards and codes and wider acceptance of water use efficiency. The result is that overall demands have decreased, resulting in lower peak and average flows. This fourth edition of M22 includes data and analysis from recent research on peak demands in residential and nonresidential buildings.

## Policy Changes in Water Rate and Connection Fees

Water rate charges and connection fees are most often based on the meter size and service line size. As a result, sizing decisions have heightened relevance, with life-cycle cost and benefit implications for both the utility and customer.

Some utilities have separated meter sizing from the connection fee process. These utilities use demand-based approaches to connection fees and infrastructure charges for new customers. These approaches are independent of the meter and service line size and enable the utility and customer to focus on the expected annual demands at the site rather than the peak instantaneous demands needed for meter sizing.

Additionally, some combined water, wastewater, and stormwater utilities have removed the stormwater charge from the base, or fixed, charge to bill properties in a tailored fashion based on their land characteristics or stormwater management practices. Thus, the fixed charge based on meter size has decreased in these utilities. This trend may have an additional benefit in the case of oversized meters. Historically, some utilities lacked motivation to downsize oversized meters to obtain better low-flow accuracy because a higher fixed charge could be obtained from the larger meter. With a smaller fixed charge based on meter size, water utilities may be motivated to replace oversized meters with an optimally sized meter.

## Growing Use of Residential Fire Sprinkler Systems

Jurisdictions have long had the opportunity to adopt requirements for residential fire sprinkler systems (RFSS) as prescribed under the National Fire Protection Association (NFPA) 13D Standard. While hundreds of local jurisdictions in the United States have done just that, the inclusion of a residential fire sprinkler requirement under the NFPA 13D Standard in the International Residential Code in 2009 spurred some state-level requirements to implement these systems. The design approaches for RFSS include many unique aspects that differ from the traditional fire connections to commercial, industrial, and multiunit residential buildings. For example, while traditional fire-protection systems for commercial and industrial buildings are usually designed under the NFPA 13 or NFPA 13R Standard and frequently require a separate and distinct line for the fire supply, RFSS designed under the NFPA 13D Standard provide several design options, including separate fire lines or a single line that provides both the domestic water supply and fire supply from the water main.

As RFSS begin to be installed in a community, the water utility often needs guidance to assist it in creating policy, procedures, and regulations for the infrastructure supplying water to these systems. In response, AWWA created a guidance report that explains the functional and design parameters for RFSS and outlines the considerations the water utility needs to set policy and procedures. The report, "Residential Fire Sprinkler Systems: Guidance for Water Utilities (2018)," is available for free download at:

https://www.awwa.org/Portals/0/AWWA/ETS/Resources/ResidentialFireSprinkler-Systems.pdf.

This report is a valuable tool for water utilities and will guide them in working with multiple stakeholders including fire departments, code officials, fire system designers, builders, and customers. The report discusses the factors around customer meter sizing; customer service line sizing, materials, and configuration; backflow protection; water rates and charges; and public relations. The report identifies appropriate references for readers to seek additional guidance. This edition of AWWA M22 details the most important elements of RFSS design guidance, particularly as related to customer service line and water meter sizing. Water utilities benefit from the information included in this edition of M22 and AWWA's RFSS guidance report to assist them in managing improved residential fire protection in their service area.

## AWWA MANUALS RELATED TO METERING AND SERVICE LINES

AWWA publishes several manuals of water supply practices that are focused on water meters and service lines. Manual M22 is one of a suite of manuals that form AWWA's core metering and service line guidance documents. A summary of AWWA manuals related to metering follows.

### M1, *Principles of Water Rates, Fees, and Charges,* Seventh Edition (2017)

M1 provides financial managers, water policymakers, and rate analysts with relevant information needed to evaluate and select water rate structures, fees, charges, and pricing policies.

### M6, *Water Meters: Selection, Installation, Testing, and Maintenance,* Fifth Edition (2012)

M6 provides a complete manual of practice for water utilities on the selection, installation, operation, and maintenance of customer water meters and also provides sample record-keeping forms, a history of water-use measurement, and the development of modern water meters. M6 is heavily illustrated with photos, diagrams, and performance requirements.

### M33, *Flowmeters in Water Supply,* Third Edition (2018)

M33 describes in detail the design and operation of commonly used flowmeters in water systems and field-testing methods.

### M36, *Water Audits and Loss Control Programs,* Fourth Edition (2016)

M36 describes how to conduct a water loss audit and provides guidance in improving accountability and economically control water and revenue losses; it includes details on production meter testing.

# REFERENCES

DeOreo, W.B.; Mayer, P.W.; Kiefer, J.; and Dziegielewski, B. 2016. "Residential End Uses of Water, Version 2." Denver: Water Research Foundation.

Mayer, P.; Buchberger, S.; Davis, S.; and Douglas, C. 2020. "Assessing Water Demand Patterns to Improve Sizing of Water Meters and Service Lines." Denver: Water Research Foundation.

Mayer, P.; DeOreo, W.; Kiefer, J.; Dziegielewski, B.; Optiz, E.; and Nelson, J. 1999. "Residential End Uses of Water." Denver: Water Research Foundation.

This page intentionally blank.

Chapter 2

# Consumer Water Demands, Trends, and Considerations

## INTRODUCTION

Peak customer demand is an important factor in many elements of the design of a water utility's production and distribution capacity and in customer metering. The maximum day demands often are used for sizing of water treatment plants,* and peak instantaneous demands, along with fire flows, are critical for sizing treated water storage and distribution systems. Another important use of peak instantaneous demands, and the main topic of this manual, is for sizing service lines and meters for individual customers.

Updates to recent end-use studies and additional data sets obtained from a broad range of utilities provide useful information for evaluating meter and service line sizing. This topic can be controversial because water rate commodity charges and connection fees are typically based on water meter size. Additionally, new development fees for sewer service can be based on water meter size and can increase disproportionately to water meter capacity. As a result, water meter sizing decisions are critical to new customers, and life-cycle cost and benefit implications from both the utility and customer perspective require thoughtful decision-making and defensible methodologies.

* Because of the high cost of providing treatment, some utilities have trended toward using peak flows over a longer period (such as 2–5 days) and relying on system storage to meet peak demands above treatment capacity. Such an approach has serious water quality implications, however. The Water Industry Database (AWWA and AwwaRF 1992) indicates an average distribution system retention time of 1.3 days and a maximum retention time of 3.0 days based on a survey of more than 800 US utilities.

Proper sizing of water lines requires anticipating all water demands, including fixtures and fire flow within the building, and then determining the smallest appropriate pipe size. Meter sizing presents additional challenges. Accurately sized water meters allow for full and accurate revenue recovery through rates as they were designed and provide satisfactory water service and full accountability for end users. Inaccurately sized water meters, either too big or too small, create a variety of financial, maintenance, and inaccurately measured demand problems.

Right-sizing water meters and service lines for modern, water-efficient buildings is aided by new data (collected through several Water Research Foundation [WRF] projects) on which to base estimates of peak instantaneous demands or flow ranges. This fourth edition of M22 includes updates for both residential and nonresidential buildings equipped with water-efficient fixtures and includes analysis from new data collected for the purpose of evaluating meter and service line sizing for multifamily and nonresidential customers.

There are seven major reference sources for sizing and typing water meters. These include four plumbing codes: Uniform Plumbing Code (UPC), National Standard Plumbing Code (NSPC), International Residential Code (IRC), and International Plumbing Code (IPC). They also include AWWA M22, AWWA M6, and local experience by individual water utilities, which develop their own criteria and tables for sizing and typing water meters to meet local water demand and pressure needs.

Four decades ago, when the first edition of M22 was written, the main approach for predicting peak water demand was called the fixture value method. This method conservatively estimated peak water demands based on required flow of then-existing plumbing fixtures and appliances. The method and resulting demand versus fixture value curves for different building types calculated worst-case instantaneous peak demand that could occur for a specific water customer. In the past 40 years, fixtures, appliances, irrigation demands, residential fire sprinkler demands, and their implications for water meter sizing have changed dramatically. Demand reductions have been driven by the national Energy Policy Act, customer education, conservation requirements and ethics, delivered water pricing, and water efficiency standards and codes.

## WATER-USE TRENDS

### Background

A common method for determining per-capita water use is to divide the sum of water used in a given period by the total number of people served. The type of customer base, climate, irrigated area, and efficiency of use thus influence per-capita water use for a water utility. In the 20th century, urban per-capita use first increased as households installed new water-using devices and fixtures and then decreased as federal plumbing codes and national standards required these products to be more water efficient, and metering and conservation pricing were enacted (Tanverakul and Lee 2015). Currently, North America is in a period of declining per-capita water use (Tanverakul and Lee 2015).

Up until 1980, urban per-capita water use in the United States increased from an annual average of about 60 gal per capita per day (gpcd) in 1920 to about 180 gpcd in the 1970s. From 1940 to 1970, per-capita water use doubled, increasing from 85 gpcd to 175 gpcd. Since the late 1980s, indoor per-capita use has leveled or declined across the United States. A Water Research Foundation (WRF) study, "North America Residential Water Usage Trends Since 1992," found the typical American household reduced its use by 12,000 gal/yr over 20 years (Coomes et al. 2010).

The Energy Policy Act of 1992 (DOE 1992) and other codes and standards have mandated reductions in toilet flush volumes, clothes washer volumes, and shower and faucet flow rates; a recent residential end-use study from the WRF documents the impact of these changes over the past 15 years. During this time, many water providers have discovered that increasing population does not necessarily result in a proportional increase in water use (DeOreo et al. 2016).

The "Residential End Uses of Water, Version 2" (DeOreo et al. 2016) provides an updated and expanded assessment of water use in single-family households across North America and presents detailed information and data about how water use has changed since WRF's often-cited 1999 "Residential End Uses of Water" was published (Mayer et al. 1999).

Water use in the nonresidential sector within a given climate zone tends to be more site-specific. This tendency is especially true in the industrial sector in which use often depends on employees, process needs, and irrigated landscape demand.

## Residential Per-Capita Use

In "Residential End Uses of Water, Version 2" (2016), average per-capita indoor daily water use was 58.6 gpcd. In the 1999 "Residential End Uses of Water," the average per-capita indoor daily water use was 69.3 gpcd. This indicates an overall 15.4% reduction in indoor per-capita usage between the two studies. This change was found to be statistically significant at the 95% confidence level. Data from these two studies were used to develop the revised residential meter-sizing methods described in this fourth edition of M22.

Additional indoor efficiency is possible and should be expected. A sample of new homes built with high-efficiency fixtures and appliances (including WaterSense-labeled toilets, faucets, showers, and high-efficiency clothes washers) had an average daily per-capita water use of 36.7 gpcd measured in research using similar methods (DeOreo et al. 2001, 2012).[†] An average indoor per-household usage level below 40 gpcd appears achievable with full saturation of high-efficiency fixtures and appliances. Several states including California, Colorado, Georgia, Texas, and Washington have mandated that fixtures sold meet WaterSense (or similar) specifications.

Things have changed a lot in 45 years. In 1980, the first maximum volume flush standard for toilets was established at 3.5 gallons per flush (gpf). Table 2-1 shows the national fixture standards for US manufacturers that took effect in January 1994 as part of the 1992 Energy Policy Act. The conversion to high-efficiency fixtures has had a major impact on annual average demand and daily peak instantaneous water demands because they affect the peak flow for each device. Before the 1992 legislation, toilet flush volumes of 5 gal and showerhead flow rates of 5–10 gpm were common.

National water efficiency programs such as USEPA WaterSense; local green building codes such as the California CalGREEN code; and states including Georgia, Texas, Colorado, and California have encouraged and mandated water efficiency beyond the

**Table 2-1** National fixture standards and specifications

| Fixture | 2019 WaterSense Specifications | 1992 Energy Policy Act Standards | 1980 Standards |
|---|---|---|---|
| Showerheads | 2.0 gpm | 2.5 gpm | 3.0 gpm |
| Lavatory and kitchen faucets | 1.5 gpm | 2.2 gpm | 3.0 gpm |
| Metering faucets | — | 0.25 gpm | — |
| Gravity, flushometer tank, and electro-mechanical hydraulic-valve type toilets | 1.28 gpf | 1.6 gpf | 3.5 gpf |
| Urinals | 0.5 gpf | 1.0 gpf | 3.5 gpf |

† New homes equipped with WaterSense toilets (1.28 gpf), showers (2.0 gpm), and faucets (1.0 gpm) and high-efficiency clothes washers.

1992 Energy Policy Act requirements. Similarly, the EnergySTAR program has resulted in more water-efficient and more energy-efficient clothes washers and dishwashers being manufactured that have provided additional indoor efficiencies. The biggest reduction in per-capita water use between the 1999 and 2016 end-use studies was measured in the clothes washer category, which fell by 35.9% from 15.0 gpcd in 1999 to 9.6 gpcd in 2016 (Mayer 2016). Over the same period, toilet use fell 23.2% from 18.5 gpcd to 14.2 gpcd.

Additional indoor conservation potential exists in the single-family sector. Daily average indoor use below 40 gpcd is achievable in the coming years through replacement of old toilets and clothes washers and as customer side leakage is reduced through automated metering leak alert programs (DeOreo and Mayer 2012).

Outdoor water use is driven by the size of the irrigated landscape, type of plant material, and weather. However, the flow rate for automated irrigation systems can be controlled by the design criteria and thus impact meter and service-line sizing to some degree. For existing sites, automated irrigation systems can often be designed based on the size of the existing water meter and service line. In these cases, flow rates are determined by the number and flow rates of the sprinkler heads (or emitters) for each station (also called *valves* or *zones*). Besides landscape irrigation, other outdoor consumptive uses include water for pools, spas, ponds, and vehicle and hardscape cleaning. Although the percent breakdown of these outdoor uses would vary widely around the country and from customer to customer, the largest percent of use on average is for landscape irrigation. In the Southwest, outdoor residential use can account for over 50 percent of the total residential annual use. More utilities are offering irrigation-specific water meters for residential use, meaning utilities will install two separate meters at homes with irrigation systems. These meters help identify the single largest residential water use and allow utilities enhanced control over water irrigation during droughts. These meters tend to improve conservation when homeowners can identify where water is used.

Per-capita indoor water use has declined significantly over the past 25 years, and the efficiency standards in place coupled with building codes that are increasingly focused on water and energy efficiency will result in further demand reductions. Outdoor water use has also decreased in some regions; however, outdoor use is more variable and subject to the impacts of the prevailing climate.

## Nonresidential Use—Improved Efficiency

In the nonresidential sector (commonly called commercial, industrial, and institutional, or CII), water demand trends are more difficult to predict because of the wide variety of land uses, manufacturing processes, and inherent site-specific differences. There will be increased efforts to ensure that water is used as efficiently as possible as demand stresses existing water supply sources, as the cost of water increases, and as incentives are provided to reduce demand. The general trend, therefore, is toward increased efficiencies of use and lower instantaneous peak demands versus historical norms.

Trends to improve nonresidential water-use efficiency include:

- increasing cooling system efficiencies;
- switching from water-cooled to air-cooled ice machines and refrigerant compressors;
- increasing water use efficiency in restrooms;
- improving the efficiencies of water-using processes and eliminating single-pass cooling;
- increasing water use efficiency in commercial kitchens;
- water reuse or recycling in car washes, commercial dishwashers, and other industrial and commercial water-using processes; and
- utility incentive programs.

# WATER USE PATTERNS

## Residential Use

Typical single-family indoor use for a standard high-efficiency home is shown in Table 2-2. Data for the standard home come from the 2016 WRF report, "Residential End Uses of Water, Version 2" (DeOreo et al. 2016). The homes in this study came from cities in the United States and Canada and reflect a reasonable mix of the current housing stock. The high-efficiency data come from a sample of homes equipped to meet the WaterSense New Home Specification, which includes 1.28 gpf toilets and high-efficiency clothes washers designed to use less than 15 gal per load (DeOreo et al. 2012). More homes are likely to use water similarly to the high-efficiency homes in the future.

## Nonresidential Use

While nonresidential water use tends to vary from site to site, ranges of use can be found for different business classes, such as schools, hospitals, offices, restaurants, etc. Per-account use and monthly peaking factors are shown in Table 2-3 for selected business classes in the East Bay Municipal Utility District (EBMUD) service area in the San Francisco (Calif.) Bay Area in the 1990s and Westminster, Colo. (a suburb of Denver) in 2017. This table is presented primarily for illustrative purposes. Average use in each category may vary considerably from community to community and from customer to customer, but there are similarities. The peaking factor is the ratio of the peak to the average month of use.

Evaluating nonresidential uses requires a comprehensive water assessment of all known water uses in the facility. Understanding how water will be used within a facility is the foundation for proper meter and line sizing. Each facility type has different water use patterns, depending on its function and use. While the equipment and processes vary widely, there are opportunities in all commercial and institutional buildings to achieve significant water savings indoors and outdoors by making improvements in operational areas. Submetering by the facility to routinely monitor component water use allows building owners and operators to more effectively understand and manage facility water use. Submetering enables a facility to quickly find and fix leaks or other unnecessary water use. The added benefit is that the facility can identify cost-effective water-use reduction opportunities and track project savings.

**Table 2-2** Per-capita residential indoor use

| | Standard | | High-Efficiency | |
|---|---|---|---|---|
| **Inside Water Use** | *percent* | *gpcd* | *percent* | *gpcd* |
| Toilet | 24 | 14.2 | 15 | 5.5 |
| Shower | 19 | 11.1 | 31 | 11.1 |
| Bath | 3 | 1.5 | 3 | 1.2 |
| Laundry | 16 | 9.6 | 11 | 4.0 |
| Dishwashing | 1 | 0.7 | 1 | 0.5 |
| Faucets | 19 | 11.1 | 17 | 6.2 |
| Leaks | 13 | 7.9 | 18 | 6.5 |
| Other/Misc. | 4 | 2.5 | 2 | 0.6 |
| **Total** | **100** | **58.6** | **100** | **35.6** |

Note: Table 2-2 was created using data from DeOreo, W.B.; Mayer, P.W.; Kiefer, J.; and Dziegielewski, B. 2016. "Residential End Uses of Water, Version 2". Denver: Water Research Foundation.

**Table 2-3** Average per-account consumption and monthly peaking factors* for selected customer types in Oakland, Calif., area and Westminster, Colo.†

| Business Type | Annual Average Consumption, gpd | Monthly Peaking Factor | Data Source |
|---|---|---|---|
| Auto Repair | 300 | 1.3 | EBMUD |
| Auto Repair | 200 | 1.6 | Westminster |
| Car Wash | 3,000 | 2.0 | EBMUD |
| Car Wash | 6,000 | 2.2 | Westminster |
| Retail | 500 | 1.3 | EBMUD |
| Retail | 1,600 | 2.6 | Westminster |
| Grocery | 800 | 1.5 | EBMUD |
| Grocery | 7,600 | 1.7 | Westminster |
| Large apartments | 3,000 | 1.3 | EBMUD |
| Multifamily (50–100 bathrooms) | 2,600 | 1.2 | Westminster |
| Hospital | 15,000 | 1.3 | EBMUD |
| Hospital | 14,100 | 1.8 | Westminster |
| School | 6,000 | 2.0 | EBMUD |
| School | 2,400 | 1.5 | Westminster |
| Hotel (with food) | 7,000 | 1.3 | EBMUD |
| Hotel | 7,700 | 1.9 | Westminster |
| Office | 1,100 | 1.7 | EBMUD |
| Office | 1,000 | 2.9 | Westminster |

* Summer (July and August) use over winter (December and January) use.
† East Bay Municipal Utility District and City of Westminster, Colo., consumption data (1997 and 2017, respectively).

## REFERENCES


AWWA and AwwaRF. 1992. *Water Industry Database*. Denver: AWWA and Awwa Research Foundation.

Coomes, P.T.; Rockaway, J.R.; and Kornstein, B. 2010. "North America Residential Water Usage Trends Since 1992." Water Research Foundation (WRF) and US Environmental Protection Agency (USEPA). Denver: WRF.

DeOreo, W.B.; Dietemann, A.; Skeel, T.; Mayer, P.W.; Lewis, D.M.; and Smith, J. 2001. "Retrofit Realities." *Journal AWWA*. 93(3):58–72.

DeOreo, W.B. and Mayer, P.W. 2012. "Insights into Declining Single-Family Residential Water Demands." *Journal AWWA*. 104(6):E383–E394.

DeOreo, W.B.; Mayer, P.W.; Kiefer, J.; and Dziegielewski, B. 2016. "Residential End Uses of Water, Version 2." Denver: WRF.

DOE (Department of Energy). 1992. National Energy Policy Act. Washington, DC: DOE.

Lee, J. and Tanverakul, S.A. 2015. "Price Elasticity of Residential Water Demand in California." *Journal of Water Supply: Research and Technology-AQUA*. 64(2): 211–218.

Mayer, P.W. 2016. "Water Research Foundation Study Documents Water Conservation Potential and More Efficiency in Households." *Journal AWWA*. 108(10):31–40.

Mayer, P.W.; DeOreo, W.B.; Opitz, E.; Kiefer, J.; Dziegielewski, B.; and Nelson, J.O. 1999. "Residential End Uses of Water." Denver: AwwaRF.

Tanverakul, S.A. and Lee, J. 2015. "Impacts of Metering on Residential Water Use in California." *Journal AWWA*. 107(2):E69–E75.

This page intentionally blank.

Chapter 3

# Demand Profiling for Optimal Meter Sizing

## ACCURATELY MEASURING CUSTOMER USAGE

Customer usage data is typically generated by attaching an electronic flow recorder directly to a customer's water meter and logging the customer usage over short increments of time. For utilities that have deployed water meters with advanced internal data storage capabilities, a mobile Automated Meter Reading (AMR) system, or a fixed-network Advanced Metering Infrastructure (AMI) system to read their meters, these technologies may also be leveraged to collect detailed consumption data. In so doing, a customer-specific demand profile is created, which accurately represents that customer's water-use characteristics.

Demand profiles generated from existing water meters are uniquely valuable because a water meter represents the most precise means to measure potable water use. Flow recorders (such as the one shown in Figure 3-1) with accompanying software accomplish their mission without interrupting the registration of the water meter and, typically, without altering the existing meter configuration. In a small number of cases, adapters are required and easily installed. The existing meter must be accurate for the data obtained by the flow recorder to be valuable. Meter accuracy specifications can be found in the AWWA M6 manual (AWWA 2012) and in the C700 series standards, which cover specific meter types.

*Source:* Courtesy of F.S. Brainard and Co.

**Figure 3-1 Meter Master EL 100 flow recorder and water meter**

Applications for customer demand profiles may be grouped into three categories: (1) meter sizing, typing, and maintenance; (2) water use audits; and (3) cost-of-service studies. While only the first category is discussed in detail in this manual, the same data gathered for meter-sizing and typing purposes has other important uses and can benefit several utility functions including distribution, metering, conservation, customer service, engineering, planning, and finance. In the case of water use audits, demand profiles assist with conservation programs, leak detection, customer service, and hydraulic modeling. For cost-of-service studies, demand profiles are used to obtain data regarding the variability of use by residential, commercial, industrial, and wholesale customer-class groups. Because the same data can support all these applications, as well as spatial variation analyses, all potential applications must be considered for which the data may be of value—either in the present or in the future—when making the decision to collect the data. For example, if a cost-of-service study or hydraulic model requires only hourly demand data, one may still choose to store the data in 10-, 30-, or 60-s increments so that the same data can be used effectively for meter sizing and maintenance programs. An example of a software program that uses demand profile data to determine appropriate meter sizing is provided in Appendix A.

Flow recorders have been used to size meters since the 1930s. Although the technology has changed a great deal since then, recorders remain the most accurate meter-sizing methodology provided that the meter being recorded is accurate per the appropriate AWWA manual or standard (e.g., M6 [AWWA 2012] or C700-series standards) and/or manufacturer standards. Mobile AMR and fixed-network AMI can be leveraged to provide usable data as well and do not require the labor that is associated with logger deployments. However, while AMR/AMI data are useful, they come with lower sample rates compared to data loggers, unless the meter and/or endpoint are specifically configured to collect data on a more frequent interval schedule—which, in turn, may affect battery life. Other methods to gauge water use, such as pipe size or fixture count, do not provide the actual usage patterns that are essential for determining the ideal meter type and size.

In general, a demand profile should accurately provide peak-usage flow rate data and the percentage and volume of water used in critical flow rate ranges. Critical flow rate ranges would include, as a minimum, flow rates below the specified accuracy range of a meter; flow rates at the crossover range in a compound meter setting; flow rates greater than the midpoint of a meter's specified flow range; and maximum flow rates. The objective is to minimize the size of the meter (and its cost) for maximum accuracy and revenue recovery without adversely affecting meter reliability, pressure levels, or fire-flow requirements. Meter maintenance costs are important to consider. It may be that a 6-in. turbo meter could better serve a customer with constant flows of 600 gpm than a 4-in.

turbo; while both would accurately measure the flows, the 6-in. turbo would experience less degradation from wear and tear. The more closely a meter is matched to a customer's usage pattern, the more water will be accurately measured and billed.

In *The Large Water Meter Handbook* (1st ed.) (Edgar 1995), Tim Edgar illustrates this potential revenue gain with the case of a 100-unit apartment building with a 4-in. turbo meter. The monthly consumption was 500,000 gal, but much of that volume was at low-flow rates. Because the turbo meter was less accurate at flow rates below 12 gpm, 15 percent of the volume went unrecorded and unbilled in both water and, as is often the case, sewer charges. The result was a revenue loss of $1,700 per year (at $3/1,000 gal for combined water and sewer). As Edgar points out, if a utility has 100 such incorrectly sized meters, those meters would cost a utility over $1 million in lost revenue over six years.

As another example, an Energy Services Company (ESCO) developed a water-loss optimization project for a large water utility in Illinois in 2018. This water utility had 404 large water meters, all of which were of the horizontal turbo type—regardless of the application. The utility initially had chosen to install horizontal turbo-type water meters because of the lower initial cost, lower maintenance costs, and ease of maintenance. In 2009, the utility did not give much consideration to the volume of nonrevenue water and to the unbilled water revenue at that time.

Shortly before commencing the water-loss optimization project, the utility had replaced the measuring chamber assemblies of 43 large horizontal turbo water meters with new chambers that were designed to accurately measure usage at low, intermediate, and high flow rates. Engineers at the ESCO identified an additional 124 large horizontal turbo-type meters that appeared to be improperly typed for the application. The engineers installed flow recorders to each of those services and analyzed the data.

Only eight of those 124 large horizontal turbo water meters were properly typed for the application. Eighteen of those 124 large horizontal turbo water meters were grossly oversized for the application. If the water meters were correctly sized, the decrease in base charge water revenue would never offset the increase in volume water revenue; there was insufficient usage at those services to recover the higher cost of the new measuring chamber assemblies that were designed to accurately measure usage at low, intermediate, and high flow rates.

Ninety-eight of the 124 large horizontal turbo water meters were improperly typed for the application. At a water rate of $2.97/100 ft$^3$ (sewer treatment was provided by a different organization and therefore not available to the utility), the utility could realize a revenue increase of at least $105,925 annually from replacing the existing horizontal turbo measuring chamber assemblies with new measuring chamber assemblies that were designed to accurately measure usage at low, intermediate, and high flow rates at those 98 services. Had the utility provided sewer treatment, the total annual benefit from large water meter typing would have been even higher.

While the most direct benefit of proper meter sizing is increased revenue, cost-of-service billing equity, and accountability, meters offer much more value than just revenue enhancement to a distribution system. Any decision made by a utility related to water usage can only be as good as the consumption data collected from meters. In general, demand profiles provide valuable data to improve distribution system design sizing, performance, and management. In addition to identifying increasing supply/customer demand water-balance accuracy and revenue, demand profiles help to:

- identify service size requirements,
- clarify meter maintenance requirements,
- define and improve water use characteristics,
- enhance customer satisfaction and awareness, and
- establish equitable and justifiable rate structures.

With increased water scarcity, conservation has also become an important industry issue. For many utilities, conservation has become the most cost-effective means to increase water resource availability. All these distribution system design, performance, and management objectives depend on the ability of a system's meters to account for usage as accurately as possible, which can only occur because of sizing and typing meters correctly.

## RECORDER DESIGN

### Theory of Operation

Demand profiles are generated with electronic flow recorders or with internal data logging using meters with electronic registers and/or AMR/AMI system endpoints, where available. The portable flow recorders discussed herein are also referred to as *demand profilers, demand recorders,* and *loggers*. These devices collect data from a meter's internal drive magnets, a meter's pointer movement, or the meter's integral encoder, digital, or analog output and then store the data that is later downloaded and analyzed. Recorders can be moved from one setting to the next with minimum effort; these devices work with standard mechanical water meters, thereby eliminating the need for special registers. A recent new recorder design by F.S. Brainard works with many electronic meters. If a magnetic or optical sensor is used, it is typically either strapped to the outside of a meter using Velcro®, cable ties, wire, or heavy-duty tape, or it is integral to an adapter located between the meter body and the existing register.

Because of adverse operating conditions (e.g., meter pits, temperature extremes, rough handling, public access), recorders should be submersible, durable, and secure. To provide extended data storage capability in remote locations, recorders should also offer substantial battery life of more than two weeks.

### Recording Methods

Data formats can be divided into three categories: pulse (or count), encoder, and analog (e.g., 4–20 mA or variable voltage). Flow recorders typically use pulse data and sometimes encoder data. Analog formats are rarely used in demand profiling. Although the analog formats are not explicitly discussed in this chapter, the following information should assist if an analog output is the only option available.

A frequently used solution for getting pulse data from existing meter installations that do not have a built-in pulse output is a magnetic flux sensor. Flow recorders using magnetic pick-ups sense the magnetic field generated by the magnetic coupling of a water meter's internal drive magnets and convert the magnetic flux change into a digital pulse that is logged into memory and later downloaded for analysis. Optical pick-up recorders sense the meter pointer passing beneath the sensor and store the signal as digital pulses to be downloaded later. Some solutions, including AMI/AMR, use an encoder interface, in which case a prior meter reading is subtracted from the current reading to calculate the volume of water for each interval. Some newer meters incorporate pulse output options that can be used for gathering usage data. In the case of pulse data, each pulse is associated with a known volume of water.

The principal advantage of a magnetic pick-up is the higher resolution of data made possible by the rotation speed of a meter's internal drive magnets. In most cases, the drive magnets inside a meter rotate much faster than the sweep hand (pointer) on the register's dial face. In small meters, the number of magnet rotations per unit of time can be as high as approximately 30 per s at 20 gpm. At this rate, the magnets are rotating 900 times as fast as the sweep hand. In the case of turbo meters, the rotation speed of the magnets can vary greatly, from approximately 800 times the speed of the sweep hand to the same speed as the sweep hand. Conversely, the advantage of using a meter's pulse or encoder output is the avoidance of possible electromagnetic interference from other equipment near a meter that a magnetic flux sensor can potentially detect. Static water meters, such as electromagnetic, ultrasonic, and fluidic oscillators, do not have rotating magnets in their respective measuring chamber assemblies, which eliminates magnetic sensors as a signal-sensing option. For these meters, a strap-on ultrasonic meter, data log stored on internal nonvolatile meter memory, or an AMR/AMI endpoint may be used for data logging. Optical and mechanical adapters enable compatibility with older gear-driven meters that preceded magnetic-drive meters.

## Magnetic Sensors

Because most mechanical water meters have the magnetic coupling directly under the register, a reliable signal can be detected by placing the sensor on the side of the register as shown in Figure 3-2. If one is present, the magnetic coupling is directly under the register in the case of all 2-in. and smaller positive displacement, singlejet, and multijet meters. With larger meters, if the magnetic coupling is not directly under the register, it is typically in the center of the turbo rotor in the middle of the flow. In this case, the magnetic sensor must be placed on the side of the meter body to be as close to the drive magnets as possible.

*Source:* Courtesy of F.S. Brainard and Co.

**Figure 3-2 Flow recorder (right) collecting and storing a demand profile from a 3-in. water meter (left)**

If the magnetic coupling is under the register, but the register has shielding on the sides, the sensor may have to be located directly on top of the register to circumvent the shield. Because the recorder's magnetic sensor is picking up the electromagnetic noise generated by a water meter, the sensor can be susceptible to picking up noise generated by other sources of electromagnetic noise, such as motors, generators, and alarm systems. The recorder's sensing circuitry should be designed to consistently pick up the magnetic signal generated by a water meter's drive magnets while minimizing the potential for picking up electromagnetic noise from other sources.

## Encoder Outputs

Water meters with an encoder output use a 2- or 3-conductor connection to enable meter reading by the utility via a touch or radio read interface device. The meter's encoder output provides the current meter reading in a text string in accordance with one of the water industry's standard protocol formats. To log the encoder output with a flow recorder, this 2- or 3-conductor cable must be temporarily connected to a recorder capable of reading the encoder output protocol. The meter readings are logged into memory, and the difference between each reading becomes the interval usage value. Because static water meter designs have neither drive magnets nor a revolving sweep hand, using a 2- or 3-conductor connection to enable meter reading or the internal nonvolatile memory of the meter are the only options available for logging static water meter usage.

## Data Storage Capacity

It is essential that a recorder or meter's internal memory has adequate data storage capacity to enable the recorder to store the total number of entries for the duration of logging. As discussed in greater detail in the "Data Storage Interval" section, flow data must be logged into memory in small time increments if accurate maximum and minimum flow rate data is to be ensured. The potential factor of difference in the observed maximum flow rate between a 10-s and a 60-s data storage interval monitoring the exact same flow is 6:1. The potential factor of difference in the observed maximum flow rate between a 10-s and a 300-s (5 min) data storage interval monitoring the exact same flow is 30:1. In other words, if a solitary flow usage of 200 gal occurred for just 10 s at a rate of 1,200 gpm, the 10-s data storage interval could detect this high flow rate of 1,200 gpm, whereas the 300-s data storage interval would observe a maximum flow rate of just 40 gpm because the 200 gal would get averaged over 5 min rather than averaged over 10 s.

This difference could have serious implications for a meter size selection. Frequently, users choose to collect and store premise water-use data for one week when assessing the size of a commercial user's meter to ensure that a representative sample of flow data is gathered. If a user is to store 10-s interval data for one week, the recorder must be able to continuously store a minimum of 60,480 intervals of data. For other applications, such as cost-of-service studies and hydraulic modeling, a smaller data storage capacity is required for meter sizing. If the same data is to be used most efficiently, however, the storage capacity should provide for high-resolution data so that the same data may be used effectively for other applications.

## Data Storage Interval

This interval is the period over which a flow recorder counts pulses before that interval's pulse count is logged into memory. The interval determines the resolution of the raw data file from which all subsequent graphs and reports are generated: the shorter the interval, the greater the detail possible in subsequent graphs and reports. Because of the quick

**Table 3-1** Meter maximum flowrate range example

| Sample Frequency | Recorded Peak GPM | Recorded Average GPM | Absolute Error % |
|---|---|---|---|
| 1 s | 7.18 | | 0.0 |
| 5 s | 7.18 | 7.10 | –1.2 |
| 10 s | 7.18 | 7.07 | –1.7 |
| 30 s | 7.18 | 6.77 | –5.7 |
| 60 s | 7.18 | 6.72 | –6.5 |
| 5 min | 7.18 | 5.97 | –16.9 |
| 30 min | 7.18 | 2.95 | –59.0 |
| 1 h | 7.18 | 2.79 | –61.2 |

*Source:* Courtesy of F.S. Brainard and Co.

nature of maximum flow events, the data log interval should be sufficiently small to detect these maximum flow events. Table 3-1 shows the analysis of meter data with various time intervals. Resampling this master data set to create new data sets at different time intervals illustrates the absolute error that would be seen with too large a data storage interval. From the table, any time interval larger than 60 s pushes the absolute error beyond 10 percent of the actual maximum flowrate.

Keeping the data storage interval small is essential for accurately determining maximum flow rates. To ensure the accurate identification of a maximum flow rate, the data storage interval cannot exceed 50 percent of the duration of a maximum flow event. For example, if an industrial customer has a particular operation that occurs just once each 30 min, lasts 30 s, and uses 500 gal of water (i.e., a demand of 1,000 gpm for 30 s), identification of the 1,000-gpm flow rate can only be assured if data is logged into memory at least once each 15 s. If the data storage interval is between 15 s and 30 s, there is an increasing likelihood that the maximum flow rate will be understated because no data storage interval is guaranteed to begin and end within the 30-s event. If the data storage interval is more than 30 s, the true maximum flow rate will not be detected; the maximum flow rate will be understated. In this example, a data storage interval of 15 s or less would show the 1,000-gpm flow rate.

There are disadvantages to making the data storage interval too small. This interval defines the size of the downloaded data file and the length of time that one can record before running out of recorder memory. The same test recorded with a 5-s interval will use six times more memory than one stored with a 30-s interval. Generally, a 10–60 s interval provides adequate detail and recording time for most applications. If one is making a long recording (multiple weeks) and a 10-s interval would use up all the logger's memory before the recording is completed, lengthen the data storage interval. Another problem with too short an interval is discussed in the following section, "Meter Pulse Resolution." Briefly, if too short an interval is used on a meter with slow-moving drive magnets (or sweep hand, in the case of optical sensors), skewing (exaggeration) of maximum and minimum flow rates can occur because there is too little data for accurate calculations. A recorder's operating instructions should identify such meters so that care is taken when selecting intervals for data presentation. Software design can also provide some correction when downloading raw data from a recorder.

### Meter Pulse Resolution

This resolution is defined as the number of pulses generated by a meter that equates to a unit of liquid measure. For magnetic pick-ups, the resolution is the number of meter magnet poles (as the internal drive magnets rotate) that equate to a unit of liquid measure. It is desirable that the internal magnets revolve as fast as possible without degrading the reliability of the meter; accordingly, the higher the number of magnet poles per unit of measure, the better. For optical pick-ups, the same considerations apply to the speed of sweep hand rotation. When logging from a meter's encoder output, the resolution will equal the digit value of the least significant wheel position included in the encoder output. The available resolution on a particular meter is due to several factors, including meter age, type, and how it was ordered from the factory. It is possible to have a meter register that displays six digits but is only programmed to transmit four of them. If using the internal memory of the meter, the data log resolution may differ from the transmitted resolution. This requires due diligence on the part of the logger operator to make sure that the correct data is being collected.

The pulse resolution (or factor) is especially important when determining maximum and minimum flow rates. The issues are similar to those discussed in the preceding section, "Data Storage Interval." Concerning maximum flow rates, if a magnet (or sweep hand) is rotating slowly, a large, short-term usage could take place without any evidence of its occurrence. For example, if a 6-in. turbo meter generates just one magnetic pulse for each 500 gal, while another 6-in. turbo generates one pulse for each 2 gal, the 250 gal usage at 1,000 gpm described in the preceding section might not even be identified by a recorder attached to the 6-in. turbo with the slow-moving magnets, while the 6-in. turbo with the fast-moving magnets would have provided 125 pulses to the recorder. Furthermore, if the recorder attached to the meter with the slow-moving magnets did detect one pulse within a 10-s interval, the recorder technology might erroneously assume that 500 gal were used during that 10-s interval, which would equate to a flow rate of 3,000 gpm. If one pulse is logged in 10 s, this is the equivalent of six pulses per min, and six pulses per min multiplied by 500 gal per pulse equals 3,000 gpm. Accordingly, a meter with fast-moving magnets can provide continuously accurate data throughout the flow ranges, whereas a meter with slow-moving magnets cannot. Likewise, using optical sensors, the faster the rotation of the sweep hand, the more accurate the resultant data. As previously noted, intelligent software design can recognize and account for such circumstances.

Minimum flow rates identify leakage rates and impact the selection of turbo versus compound meters in larger flow rate applications. To ensure the accurate identification of minimum flow rates, as with maximum flow rates, a user must know which meters have slow-moving drive magnets. For example, if a meter's magnets are providing just one pulse for each 20 gal, and the current flow rate is a steady rate of 5 gpm, only one pulse will be generated each 4 min. If one observes the data in time increments smaller than once each 4 min, the flow rate will appear to vary between zero and some amount greater than the actual flow rate of 5 gpm. As an illustration, if a 1-min data storage interval is used, the flow rate will appear to equal zero for 3 of each 4 min, and 20 gpm for 1 of each 4 min, because each pulse, equaling 20 gal apiece, will appear just once each 4 min when a steady flow rate of 5 gpm is occurring. If a 4-min time interval is used to observe the data, it will appear as if a steady flow rate of 5 gpm is occurring.

The two graphs shown in Figures 3-3a and 3-3b represent the scenario just described. Both graphs were generated from the same data, but the time increments used to view the data are 4 min and 1 min, respectively. Each pulse from the meter equals 20 gal, and they were spaced 4 min apart (except during the initial interval shown). Software design can help by evaluating the data to determine the likelihood that raw pulse data should be averaged over longer periods of time.

Unless someone is at the meter site watching the meter at the time of the event, it is not possible to know with certainty whether a periodic use of 20 gal is occurring or a steady flow rate of 5 gpm is occurring. If each pulse from the meter equaled a smaller amount of water, such as one gallon, the recorder technology would be able to consistently display more accurate data.

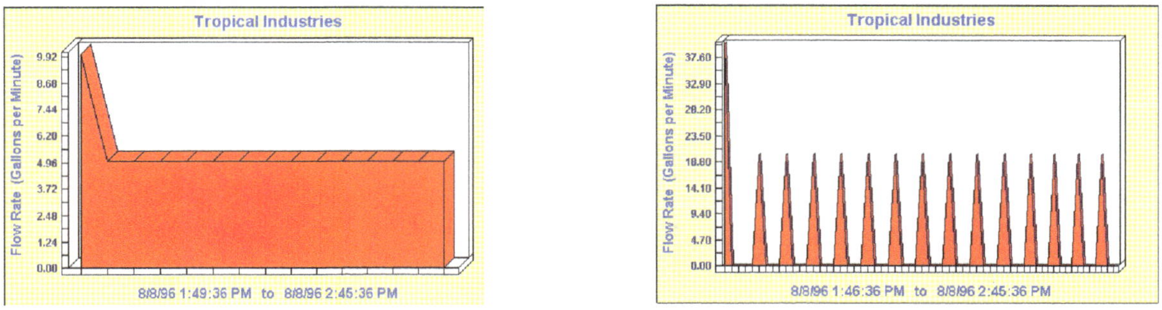


(a) Storage interval of 4 min

(b) Storage interval of 1 min

*Source:* Courtesy of F.S. Brainard and Co.

**Figure 3-3 Depiction of 4-min and 1-min data storage intervals**

*Source:* Courtesy of F.S. Brainard and Co.

**Figure 3-4 Compound adapter and sensor**

The key to getting accurate flow data is generating a sufficient number of pulses per time interval while keeping the storage interval as small as possible. In the case of magnetic pulses, most 2-in. and smaller positive displacement and some singlejet and multijet meters provide a good pulse resolution, such that the data can be observed in time increments as small as 10 s. Because some turbo meters have magnets that rotate relatively slowly, low flow rates are difficult to accurately observe.

Accordingly, adapters that increase the magnetic pulse resolution can be useful in determining accurate flow rate data because the flow data may be accurately logged and viewed in smaller time increments. One compound adapter, for example, can increase the resolution by a factor of 12 on various compound- and turbo-type meters (Figures 3-4 and 3-5). The adapter installs easily and enables the existing register to continue functioning normally. The same considerations also apply when using optical sensors.

*Source:* Courtesy of F.S. Brainard and Co.

**Figure 3-5 Recording demand profiles from large compound meters may require an adapter.**

# RECORDING DATA

## Length of Record

As previously discussed, many recorder users choose to store data from commercial sites for one week because certain high-rate water uses, e.g., a manufacturing operation, may only occur on a particular day each week. If possible, water usage should be discussed with a customer before storing data to ensure that the duration of the recording period is sufficient to get a representative sample of the full range of flow data. In the case of multitenant residential sites or hotels/motels, 24 h of data may be sufficient if the data is collected during hot weather for residential sites and during high occupancy for hotels/motels. Understanding a user's water use characteristics is important when selecting the optimum length of the data storage period. Experience with different types of users over time will also provide an indication as to the optimum record length for different classes of users.

## Customer Water-Use Patterns

Data should be recorded during a period in which the user experiences typical peak, average, and minimum flow rates and for a duration sufficient to capture those rates. For example, data should not be recorded at a school or factory during a vacation period. Similarly, as previously mentioned, utility personnel may want to record data for at least a week at an industrial site if there is evidence that the customer performs different operations on different days of the week. Seasonal cycles are as important to consider as weekly ones. Weather at different times of the year may substantially alter demand patterns, especially if outdoor irrigation use is recorded through the same meter. If a customer uses a lot more water on a hot summer day, data on such a day should be recorded to capture peak flow data for meter typing, sizing, and accuracy evaluations.

Potential changes in demand patterns should be anticipated. At a residential development, the number of additional units currently under construction should be considered. Commercial lease space can have a high rate of turnover. A bottling company could be replaced by a warehousing or distribution company with substantially lower water usage. An elementary school may be converted into an apartment complex. If the appropriate type of meter is not installed and/or resized to reflect this change, the new user could be the beneficiary of a lot of free water that is inaccurately measured at low flow rates. This results in an increased volume of apparent water loss and reduces billable revenue.

## Meter Accuracy

A flow recorder is generally only as accurate as the meter from which it is recording. When a flow recorder is used, it is typically assumed that the meter being monitored is accurate. A flow recorder cannot determine meter accuracy, but it can help determine the proper meter type and size for an existing setting. Because a flow recorder is only as accurate as the meter to which it is attached, routine meter testing is important when using recorders to determine the appropriate meter size. Since most meter inaccuracy involves under-registration of usage, a flow recorder on an under-registering meter can result in the selection of an undersized meter.

Ideally, a meter should be tested for accuracy—and repaired/recalibrated if testing indicates that it is not accurate—before recording data for meter-sizing purposes. As discussed in the section, "Meter Maintenance Considerations," a demand profile performed in conjunction with a flow test may indicate that all the flow is occurring in an accurate range of the meter, even though the meter is not accurate throughout the flow ranges. If this is the case, the meter does not need to be repaired/recalibrated because no revenue is being lost.

Flow recorders should be considered a valuable companion tool as part of a meter test program. As referred to previously, a flow recorder can identify the percentage of flow in low-, medium-, and high-flow ranges. With this information, testing can be focused on the ranges in which most of the usage is occurring, and unnecessary and costly repairs can be avoided. If a flow recorder indicates that all the flow at an oil refinery or brewery is occurring in a high-flow range, whether the meter is accurate at low and medium flow rates is not relevant.

## Verifying Data Accuracy

A principal advantage of recording flow data directly from water meters rather than using alternative technologies, such as ultrasonic devices, is that the resultant flow data is based on and may be verified against the meter's registration. Graphs and reports generated from the data may be used with confidence because a water meter is the most accurate and reliable means to measure potable water use. However, if the accuracy of the data generated with a flow recorder is not verified by comparing the total volume observed by the flow recorder to the total volume registered by the water meter itself during the data storage period, this key advantage is lost.

Verification of data accuracy is critical and is accomplished by (a) requiring the user to enter the beginning and ending meter readings when downloading data, and (b) having an accurate meter magnetic-pulse factor database so that the total volume registered by the meter may be compared to the total volume registered by the flow recorder. This procedure also requires the operator to take care when making a record of the meter readings so that the numbers are accurate and include digits down to the decimal. To read a meter down to the decimal, a digit for all rotating dials and painted-on zeros must be read.

The sample software screen shown in Figure 3-6 requires the user to compare the meter's register volume to the flow recorder's observed volume. The numbers should either be extremely close or differ by an explicable margin. The software calculates register volume by subtracting the meter's beginning register reading from the ending register reading. This example is based on a magnetic pick-up. It calculates the recorder volume by multiplying the total magnetic pulse count for the entire recording period by the magnetic pulse factor for that meter in the software's database. An explicable difference between the two total volumes would include differences due to change gears used in some meters

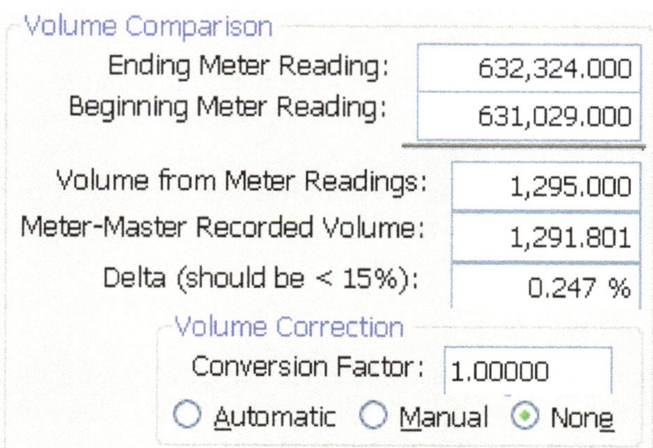


*Source:* Courtesy of F.S. Brainard and Co.

Figure 3-6 Verification of data accuracy

for calibration purposes. Because the change gears are used to speed up and slow down the register to match the activity in the meter's chamber, the recorder's volume could differ from the register's volume by as much as 15 percent, even though both the meter and the recorder may have functioned 100 percent accurately. The software screen shown in Figure 3-6 includes an automatic data conversion factor option so that the recorder's volume can automatically be calibrated to match the meter's volume 100 percent. These changes should only be performed by professionals with experience recording and analyzing water meter data.

## Presentation Options

Data can be presented in many formats and styles, as shown in Figure 3-7. Generally, the flow recorder results should provide the following analysis:

- Seasonality volume trend
  - Illustrates the proper choice of data logging study time
- Volume data over the study period
- Max/Avg/Min flow rate
- Volume transported at different flow rates
- Comparison across meter types
  - Meter size
  - Service type
    - Example: Healthcare, industrial, or bulk transport

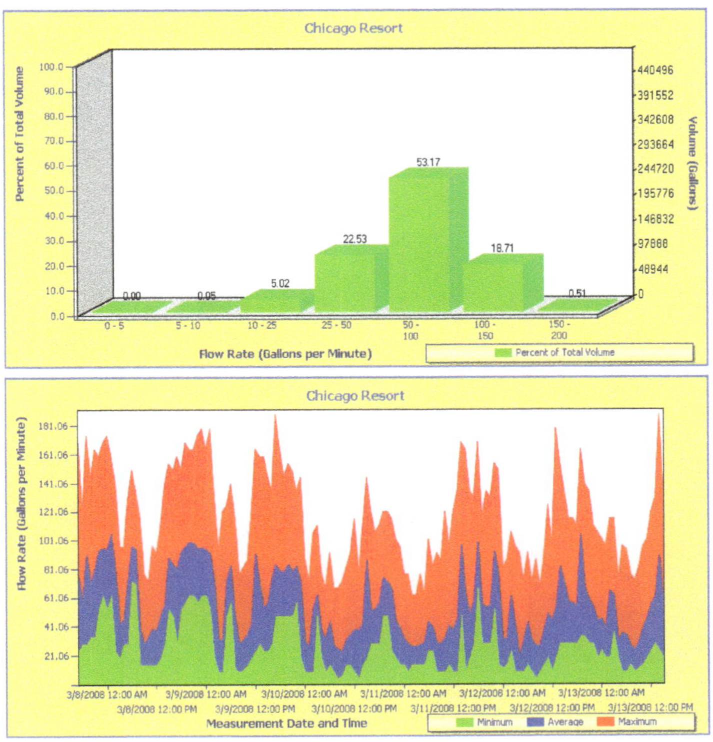


*Source:* Courtesy of F.S. Brainard and Co.

**Figure 3-7 Data presentation options**

# USING CUSTOMER USAGE DATA

## Summary of Meter Sizing Benefits

The use of demand profiles for meter sizing applies to all users except single-family residential. Standard meter size and water-use patterns characterize single-family residential customers. With users other than single-family residential, each customer generates a unique demand profile, and the meter should be sized accordingly. Although generic demand data can be developed for various customer class groups based on demographic and business information, the cost of gathering customer-specific demand data is minimal when compared to the revenue and community relations benefits associated with maximizing meter accuracy and water use accountability.

Multifamily residential meters (e.g., apartment buildings) are the most consistently oversized meters because of both traditional fixture count methods of sizing and the advent of efficient, low-flow-rate fixtures. The graph displayed in Figure 3-8 is from an 8-in. wholesale connection serving a residential community. Although the specified accuracy range for an 8-in. turbo meter is approximately 40–3,500 gpm, the flow rate never exceeded 40 gpm at this site. The customer received much free water. Replacement of the meter with one that is properly sized will result in both greater meter accuracy and higher and more accurate billed water revenue. The revenue gain is further increased by higher sewer revenues whenever the sewer charge is a function of water usage.

Before downsizing a meter at an apartment complex or any potentially large variable-use facility, the utility should make sure that the meter does not serve any fire-suppression systems. If the meter does serve such a system, the meter may not be a good candidate for downsizing because of fire flow requirements. Moreover, either a UL-listed (Underwriters' Laboratories) or FM-approved (FM Global/Factory Mutual Insurance Company) meter should be installed in this setting—preferably a fire service or comparable type of meter that is designed to accurately register usage across a wide range of flows, especially at flow rates below 7 gpm.

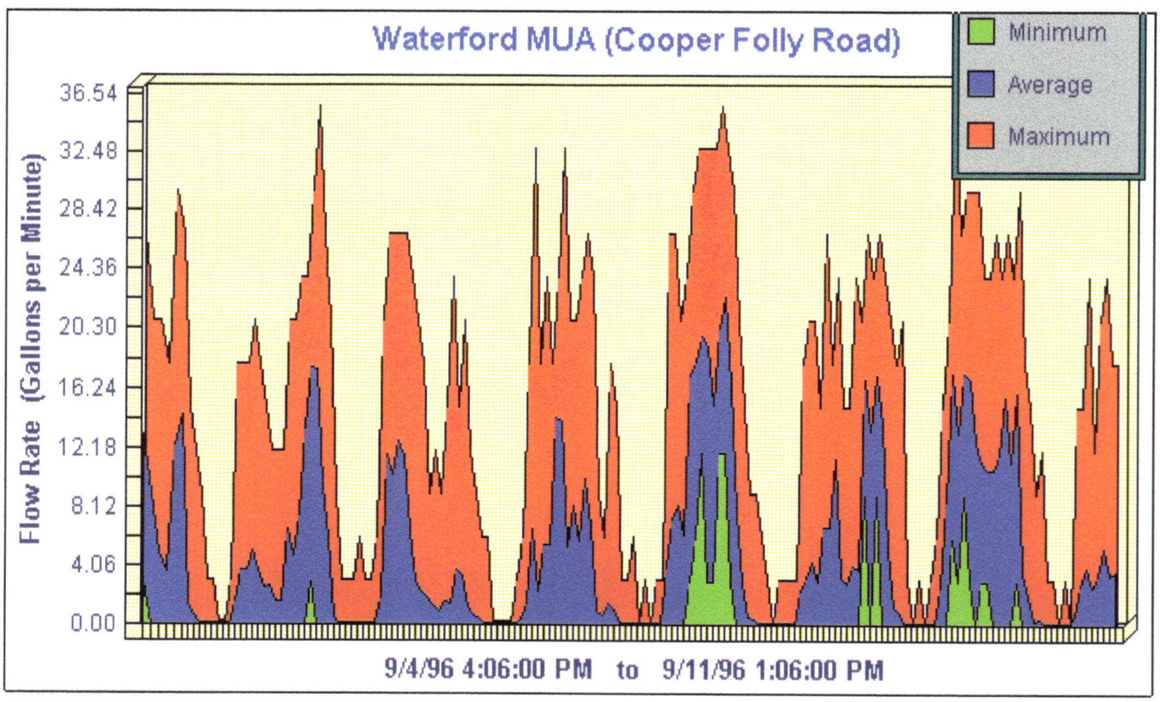


*Source:* Courtesy of F.S. Brainard and Co.

**Figure 3-8 Flow data from an apartment building served by oversized 8-in. turbo meter**

Accurate meter sizing has positive residual effects with other programs mentioned in this manual. For example, a cost-of-service study in support of a rate-structure design can only be fair and equitable if all the sample sites have properly sized meters generating accurate usage data. Modern nonrevenue water management methods rely on accurate meter sizing. Customer-side leak-detection efforts are undermined if a meter is oversized—because low flows are not detected—and the meter's pulse resolution is less than it would be with a smaller meter. Similarly, hydraulic models, conservation efforts, and other programs all benefit from accurate use registration, which depends on correct meter sizing.

### Meter Maintenance Considerations

Another related use of customer usage data is meter maintenance programs—especially the larger ones. Some utilities consider demand profiles when making meter test, repair, and/or replacement decisions because the demand data enables the utility to perform an accurate cost–benefit analysis of these three maintenance options on a case-by-case basis. For example, if a 10-in. turbo meter tests 100 percent accurate in a high-flow range, 90 percent accurate in a medium-flow range, and 80 percent accurate in a low-flow range, the conventional wisdom would average the three accuracies, which would equal 90 percent, and recommend repair. If a demand profile indicates that the flow rate never drops below 1,000 gpm, the in-service meter accuracy for the subject application would equal 100 percent because all flow is occurring in a high-flow range. With the advantage of a demand profile, costly and unnecessary service interruption and repair costs can be avoided. Proper check-valve operation in a compound meter setting can also be evaluated by ensuring that the turbo side does not move unless the bypass chamber side exceeds a specified flow rate.

Water meters, like any piece of machinery, have optimum performance ranges, and projected test requirements can be related to a user's demand profile. If a 4-in. meter is constantly being driven at a flow rate close to its high-end performance rating, more frequent repair requirements can be anticipated.

## REFERENCES

American Water Works Association (AWWA). 2012 (5th ed.). M6, *Water Meters—Selection, Installation, Testing, and Maintenance*. Denver: AWWA.

Edgar, T. 1995 (1st ed.). *Large Water Meter Handbook*. Dillsboro, N.C.: Flow Measurement Publishing.

This page intentionally blank.

Chapter 4

# Estimating Peak Water Demands for Service Lines and Meters

## INTRODUCTION

### Hunter's Curve

To properly size service lines and water meters, it is essential to know the peak demand. Chapter 3 discussed approaches for measuring peak demand from existing customers. For customers requiring a new service connection and meter, the peak demand must be calculated based on building characteristics. This problem was investigated during the 1920s and 1930s by Dr. Roy Hunter, a research physicist at the National Bureau of Standards. Recognizing that water use is a random process, Hunter (1940) applied probability principles to predict peak demand as a function of the number of fixtures in a building.

Hunter formulated his approach using the binomial distribution, a probability model for random trials with two outcomes. In the case of a plumbing fixture, it is either "on" or "off." Hunter focused on the three groups of fixtures that exerted the greatest load-bearing effect on plumbing systems in the 1930s—namely, the flush-valve toilet, the flush-tank toilet, and the bathtub. Hunter reasoned that the probability a fixture is on (running water) depended on the ratio of two factors: (1) the average duration of flow when the fixture is operating and (2) the average time between successive operations.

Estimates for time between successive operations were based on the key assumption that all fixtures experienced *congested service*. Hunter defined congestion as the maximum practical rate at which fixtures can be used continuously in actual service. This implies there are users waiting in queue for each fixture so that when one user is finished, another immediately engages the fixture. In this manner, each fixture is continuously occupied, thus maximizing the probability that the fixture is running water. Such a scene is common for restroom use during breaks at a concert or halftime at a sporting event. From observations made at a few locations, Hunter determined the probability of use for the three key fixtures was 0.030 for a flush-valve toilet, 0.200 for a flush-tank toilet, and 0.067 for a bathtub.

With these three probability parameters, it was relatively straightforward for Hunter to develop individual design curves showing peak water demand versus fixture count for each of the three fixture groups. The challenge, then, was how to combine the individual curves into a single design graph that reflected the peak water demand in a building where all fixture groups could operate simultaneously. Hunter recognized that it was not legitimate to simply add the peak flow from each individual design curve. Hunter's ingenious solution was to introduce the *fixture unit* concept, effectively collapsing the individual fixture design curves into a single composite graph giving the peak demand (gpm) versus total fixture units for any combination of indoor end uses. The final result, called "Hunter's curve," is the basis for plumbing codes in the United States and across the globe. Before this historic development, there was no uniform method among plumbing codes to estimate the peak water demand in a building. Hunter's work established the standard for satisfactory service.

It is beyond the scope of this manual to replicate Hunter's derivation of fixture units (interested readers should consult Hunter's 1940 report). The key point is that fixture units provided the common currency that enabled any assortment of indoor plumbing fixtures to be combined and expressed as a single metric, which in turn mapped directly to a single estimate of the peak demand in the building. Hunter's curve gives the 99th percentile of peak period water demand. This means there is only a 1 percent chance that the actual peak period demand will exceed the load estimated from Hunter's design chart. Thus, in a properly designed system, peak flows are expected to exceed Hunter's curve 1 percent of the time and to fall below it 99 percent of the time during the period of peak use.

Hunter's iconic curve reflected good engineering judgment for water use habits in the mid-1900s. By 1970, however, there was a growing consensus among practicing engineers that Hunter's method led to overdesign of the water supply system in new buildings (USNCCIB 1974; Benazzi et al. 1976; Steele 1982; Jack and Whorlow 2017; Hobbs et al. 2019). There were two main reasons for this. First, the assumption of congested service often did not hold. Second, flow rates based on plumbing fixtures from the 1930s did not apply to the new generation of water-conserving fixtures. The disparity between peak water demands observed in the field and predicted by Hunter's curve widened with passage of the 1992 federal Energy Policy Act (EPAct) and adoption of other green plumbing codes.

## Changes to Hunter's Curve

The profession has pursued two different paths to address these issues. One route, taken by plumbing-code committees, follows Hunter's original formulation. The basic strategy is to list additional building-occupancy types along with their corresponding modified fixture-unit values. Often these modifications stem from engineering judgment rather than observable data, resulting in discrepancies among fixture-unit values posted in various plumbing codes (Cole 2012).

The other course of action involves collecting water use readings at various buildings to generate new empirical peak demand curves. For instance, previous editions of AWWA M22 (1975, 2004, 2014) include a family of peak demand curves for different end-use categories: residential suburbs; commercial and institutional; and residential apartments and condominiums. In a subtle departure from Hunter's fixture unit concept, measured peak flows are plotted against the sum of all fixture flow rates (known as the combined *fixture value*) in a given building. A fixture value is defined as "the maximum discharge in gpm of a fixture when it is used individually without the intervention of any fixture at 60 psi" (AWWA 2004). The chief shortcomings of these empirical M22 curves are the small sample size and outdated water use readings. The field measurements were collected at just a few buildings in the United States and Canada in the early 1970s, years before the 1992 EPAct legislation. Refer to Chapters 2 and 3 for further discussion of US legislation and its impact on water fixture demands.

Besides the AWWA M22 manual, organizations such as the International Association of Plumbing and Mechanical Officials (IAPMO) and the International Code Council (ICC) publish the Universal Plumbing Code (UPC) and International Plumbing Code (IPC), with curves of peak demand versus fixture values or fixture units. The UPC, the IPC, and M22 are standard references for sizing service lines and water meters in North America. Although the UPC and the IPC use the same design curves, there are important differences between their approaches for estimating peak demand. Fixture units are assigned differently in the UPC and the IPC, and neither is equivalent to the fixture value calculation presented in M22.

In one area, the three methods agree: the UPC, the IPC, and M22 all include separate demand curves for buildings equipped predominantly with (1) flush-tank toilets or (2) flush-valve toilets. This important distinction between tank and valve toilets was also noted by Hunter. For design purposes, Hunter provided two curves in situations where the fixture unit sum is less than 1,000: Curve 1 where flush-valve toilets predominate and Curve 2 where flush-tank toilets predominate. When the fixture unit sum exceeds 1,000 on Hunter's chart, both curves merge into a single line for estimating peak water demand irrespective of toilet type.

The need for two curves in small-scale settings arises from the different hydraulic operation of flush-valve and flush-tank toilets. Valve-operated fixtures lack a reservoir of stored water and, when used, produce a brief but high demand for water on the premise plumbing. For instance, a single flush of a valve fixture may last only a few seconds while drawing up to 25 gpm. In contrast, tank-type toilets include a reservoir of stored water that is used to produce the flush. It is the refilling of this tank that produces the demand for water on the plumbing system. The refilling process may take 30–60 s with a draw in the range of 2–3 gpm. This distinction generally divides residential (flush tank) buildings from nonresidential (flush valve) buildings. In practice, small nonresidential buildings often install flush-tank toilets to reduce the meter size requirement and save money on connection fee charges.

## Recent Studies of Peak Indoor Water Demand

Since publication in 2014 of the third edition of M22, a pair of independent but complementary investigations of peak indoor water use have been completed. The first study, performed from 2011 to 2017 under the auspices of IAPMO, the American Society of Plumbing Engineers (ASPE), and the Water Quality Association (WQA), developed a probabilistic model to estimate peak demands in residential buildings with water-conserving fixtures. The second study, performed from 2016 to 2019 under the auspices of the Water Research Foundation (WRF), compared high-resolution water-use readings against the rated capacity of existing meters at a wide variety of buildings, each with water-conserving fixtures. Both studies produced valuable information and insights that are used to update and revise the peak flow charts in this 4th edition of M22.

**First study (2011 to 2017).** IAPMO and ASPE in collaboration with WQA convened a task group to review methods for estimating peak demands in buildings. This initiative was in response to the increased use of water-conserving plumbing fixtures, fixture fittings, and appliances and the subsequent decreased demand for water in commercial buildings and private residences. The charge of the task group was to develop a probability model for predicting peak indoor water demand at single- and multifamily dwellings having water-conserving plumbing fixtures. Six types of indoor household fixtures were considered: bathtubs, clothes washers, dishwashers, faucets (kitchen and bathroom), showers, and toilets. The task group analyzed high-resolution (10-s) indoor water demands collected for two weeks at over 1,000 single-family homes in 62 US cities. The analysis led to the Water Demand Calculator (WDC), discussed later in this chapter as the recommended method for estimating peak demand in residential buildings. The WDC provides an estimate of the 99th percentile of peak indoor water use in single-family homes and multifamily apartments, based on the number and type of water fixtures in each household as specified by the user. In contrast to previous peak flow estimators, the WDC does not use fixture units or fixture values. Instead, the WDC uses "p-values" which represent the probability of fixture use during the hour of peak water demand. The WDC was incorporated into the 2018 Uniform Plumbing Code (IAPMO 2018b). It is available as an Excel spreadsheet from IAPMO's website (www.iapmo.org/water-demand-calculator/) and as a smartphone app. The WDC is the recommended method for estimating peak water demand in residential buildings and is demonstrated later in this chapter.

**Second study (2016 to 2019).** The Water Research Foundation sponsored WRF project 4689, "Assessing Water Demand Patterns to Improve Sizing of Water Meters and Service Lines" (Mayer et al. 2020). The research project had two primary goals. The first was to collect peak demand flow data, fixture counts, and other pertinent information from large buildings with water-efficient plumbing fixtures (80 percent had 1–2-in. meters; 18 percent had a 3-in. meter; and 2 percent had a 4-in. meter). The second goal was to compare the measured peak water demand against the rated meter capacity and service line size at each building. Over 140 buildings were monitored by participating water utilities in three cities: Scottsdale, Ariz.; Denver, Colo.; and Westminster, Colo. The research team correlated high-resolution (10-s) flow data (from Scottsdale and Denver) and hourly flow data (from Westminster) to water meter sizes, building use, and the numbers and types of fixtures in multifamily and nonresidential buildings. The expected peak water use for buildings was estimated according to several different building codes and compared against observed water demands. In addition, the observed instantaneous peak water demands from residential buildings provided an independent data set to assess and verify the performance of peak flow predictions from the WDC. Among the key findings from this study, two are particularly relevant to M22:

- Measured peak flows in the buildings were well below the rated meter capacity.
- Demands from meter sizing design curves usually greatly exceeded measured flows.

# ESTIMATING PEAK DEMAND FOR NONRESIDENTIAL BUILDINGS

## Updated Demand Curves

This section provides guidelines for using the M22 approach to determine peak water demands in nonresidential buildings. The peak flow data collected during WRF project 4689 (Mayer et al. 2020) were used to update the existing water demand curves in the third edition of M22. Figure 4-1 shows more than 50,000 data points representing the top 1 percent of flows measured in 98 buildings for which fixture values were available at study sites in Denver, Scottsdale, and Westminster. The orange data points show

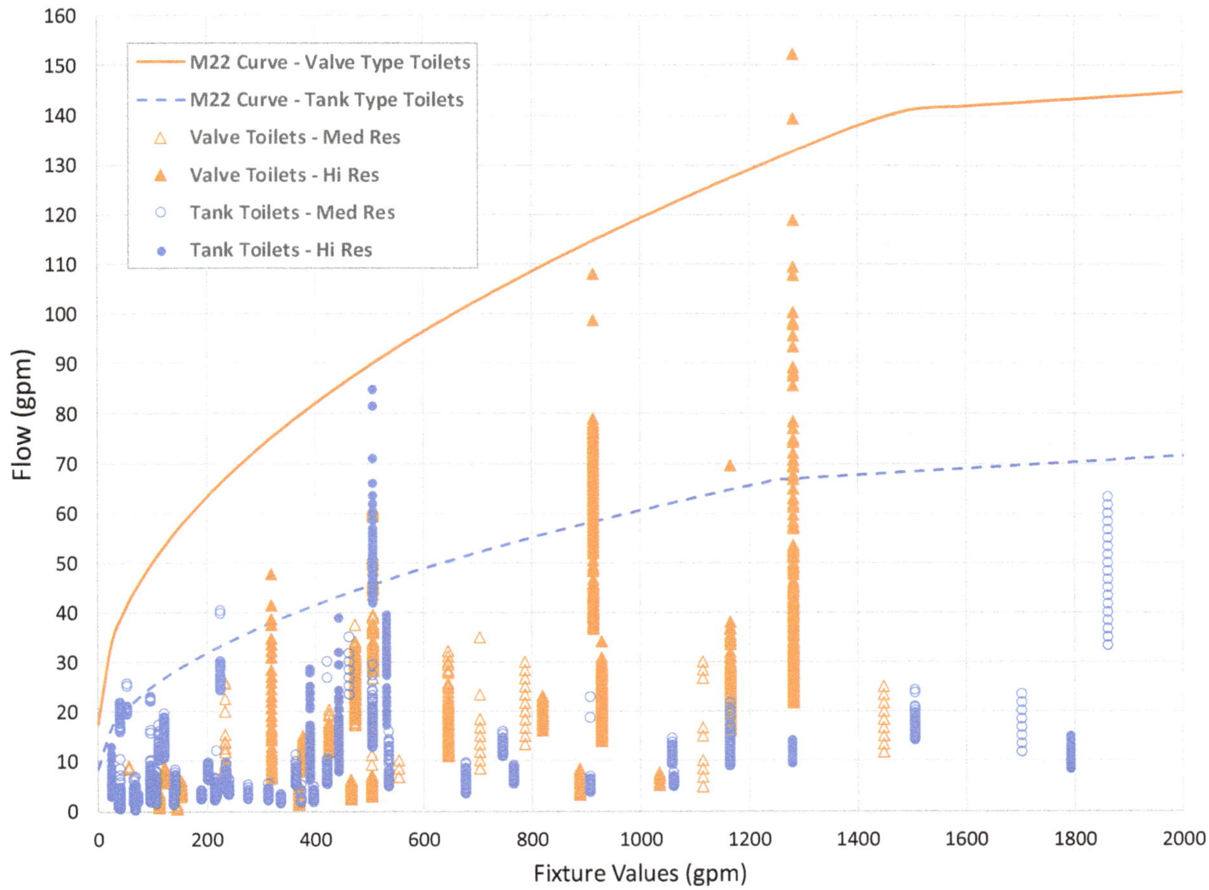


*Source:* Mayer et al. 2020. Reprinted with permission. © The Water Research Foundation.

**Figure 4-1 Design curves from third edition of M22 (2014) for valve-type and tank-type toilets plotted with top 1 percent of flows from 98 study sites in Denver, Scottsdale, and Westminster**

demands measured in 34 buildings with predominantly valve-type toilets; the blue data points depict demands measured in 64 buildings with predominantly tank-type toilets. In addition, Figure 4-1 includes the water demand curves for flush-valve toilets (solid orange line) and flush-tank toilets (blue dashed line) from the third edition (2014) of M22.

As illustrated in Figure 4-1, the highest 1 percent of measured flows seldom exceed the 2014 M22 design curves and only at a few buildings.

The flow data depicted in Figure 4-1 were used to compute a new reference peak demand for each of the 98 buildings. The new reference peak demand for each building was found as the sum of the average plus one standard deviation, using only data from the top 1 percent of the measured flows at each site. The standard deviation was included to provide an added degree of conservatism. In a few cases, the curves from the third edition of M22 were used to fill in gaps where the observed flow data from WRF project 4689 (Mayer et al. 2020) were sparse. These extrapolations occurred near the origin and at the upper end of the curves (i.e., the regions of less than 40 fixture values and greater than 7,000 fixture values).

A logarithmic regression was developed to find the best fit trend line describing reference peak demands versus building fixture values. This regression was done for 64 buildings with predominantly tank systems and for 34 buildings with predominantly flush-valve systems. Both regression lines were then adjusted upward by averaging the

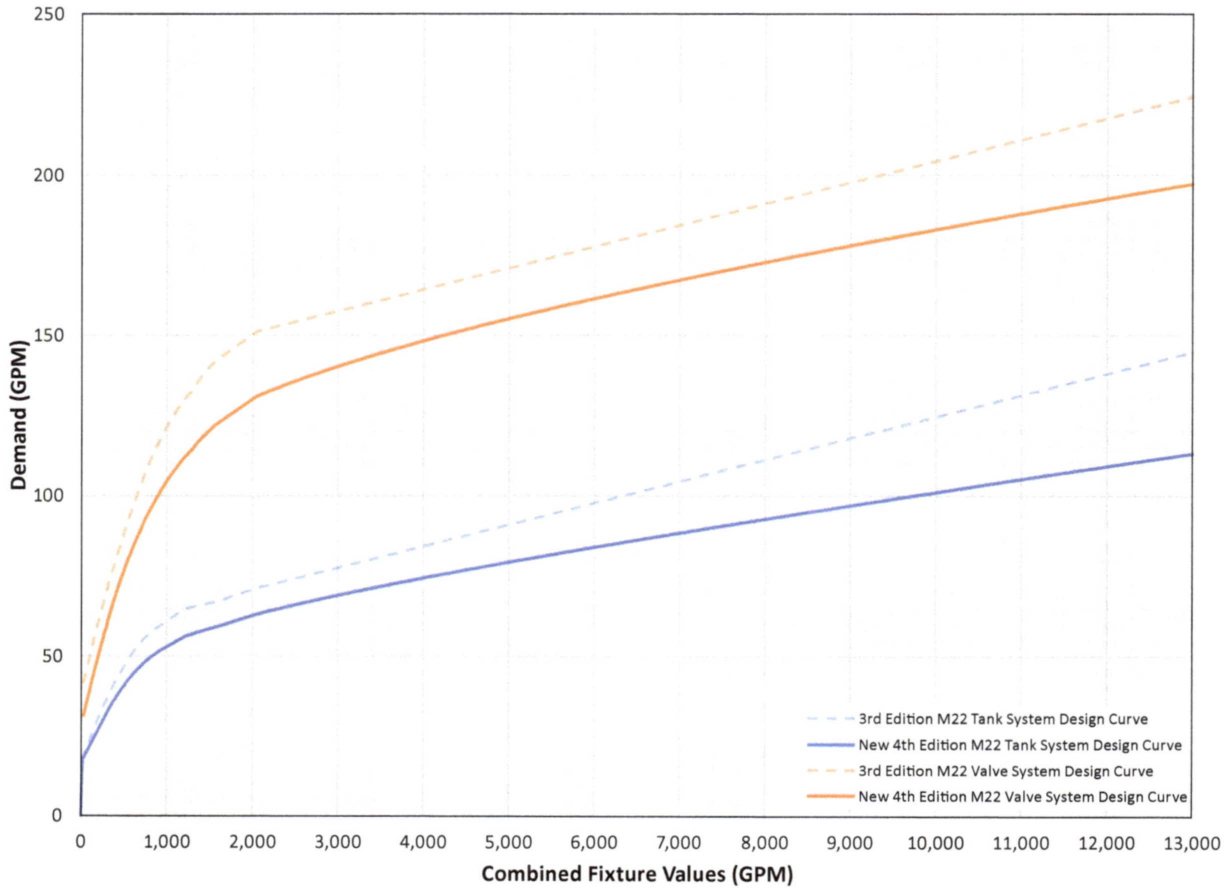


**Figure 4-2 Peak water demand versus fixture values for nonresidential buildings. (Orange lines are for flush-valve systems, and blue lines are for flush-tank systems. Dashed lines are the old design curves from the third edition of M22. Solid lines are the new updated design curves for the fourth edition of M22.)**

new trend lines with the corresponding historical M22 design curves. This final step was important near the origin of the chart where the new design curves converge with the historical design curves. The new design curves are plotted as the bold solid lines in Figures 4-2 and 4-3, along with the historical M22 design curves which appear as dashed lines. The new design curves reduce the estimated peak flow from the historical M22 curves by an average of about 16 percent for flush-tank systems and 12 percent for flush-valve systems.

Similar to the historical M22 curves, the new peak demand curves are not linear. The reason for this pattern is that the probability of all fixtures operating at one time diminishes as the number of fixtures or appliances increases. The fixture value in different types of customers will have different impacts on peak demands. A hotel may have an entirely different demand pattern than an office, school, or restaurant.

## Understanding the Fixture Value Approach

This procedure is based on an updated fixture value approach originally created in the 1975 M22 Manual. There are multiple types of water demands that must be considered for the peak flow calculation, including domestic or indoor demand, continuous demand due to hose bibbs and cooling loads, landscape irrigation demand, and fire sprinkler demand.

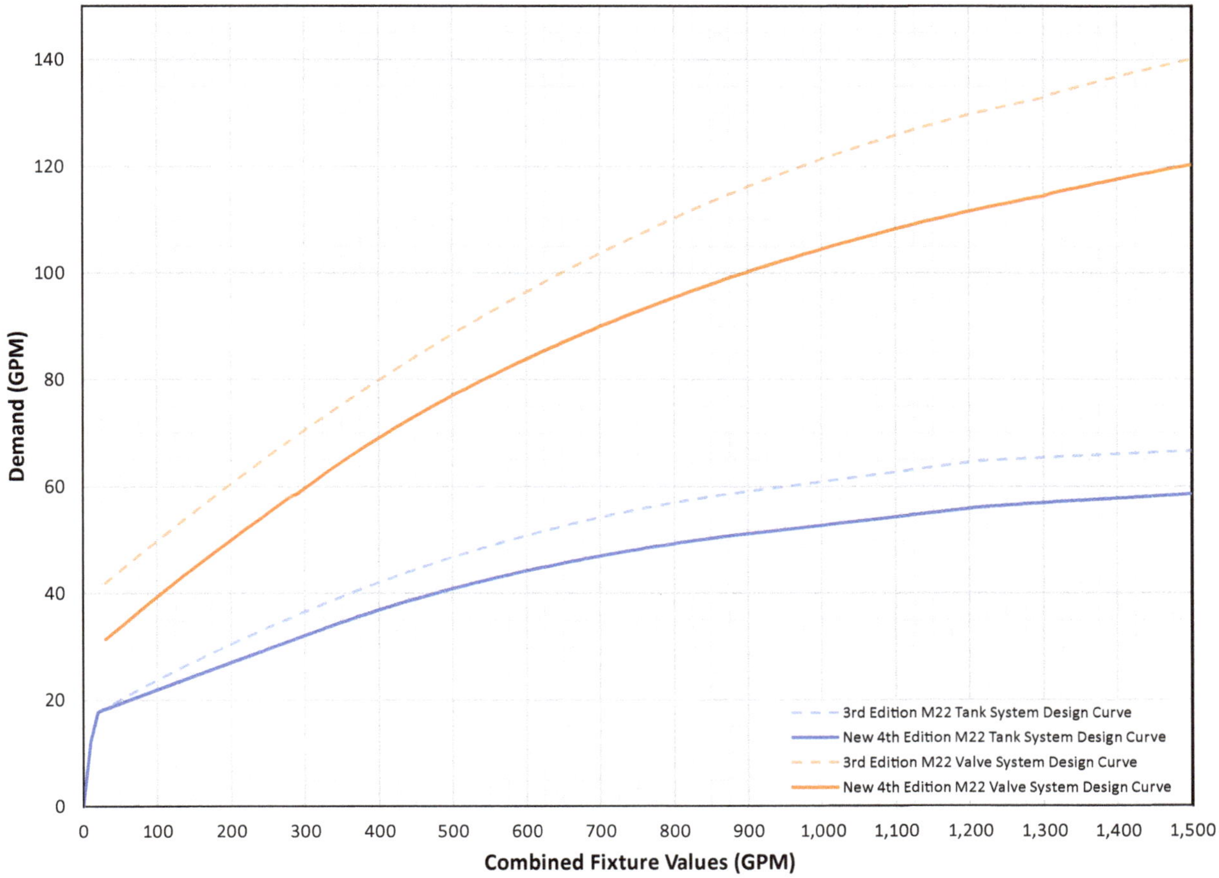


**Figure 4-3 Peak water demand versus fixture values for nonresidential buildings (scale is enlarged from Figure 4-2)**

Where base-load continuous demands are present or intermittent irrigation demands occur, these demands must be considered in meter and service-line sizing. Irrigation demands that occur concurrently with peak domestic demands must be added to the indoor domestic demands. Where irrigation demands normally occur during off-peak domestic-use times, the controlling peak demand should be the larger of the two. Where continuous demands (e.g., hose bibbs, cooling loads) occur, these must also be added to the peak domestic demands. For additional information on landscape irrigation and fire sprinkler demand, see the section titled, "Estimating Peak Demand for Other Cases," toward the end of this chapter.

The following steps and information are M22 guidelines for estimating peak indoor water demand in nonresidential buildings from domestic use of plumbing fixtures and appliances. Nonresidential building meters should be sized using fixture values. As defined previously, a fixture value is the maximum discharge of a fixture when it is used individually without the intervention of any other fixture. The fixture value serves as a simple independent variable against which measured peak demands can be plotted to develop an empirical design curve. The assumed maximum discharge for each fixture is based on the manufacturer's listed maximum flow rate according to the fixture's standard testing pressure. Suggested manufacturer peak flow rates and other specifications are listed in Table 4-1. Many fixtures operate over a range of pressures and, depending on the operating pressure at the fixture, the actual flow rate can fluctuate. Table 4-1 also lists the water flow rate when the fixture is operating without the interference of other fixtures at 60 psi. Values for fixtures or appliances not listed in Table 4-1 should be obtained from the manufacturer.

## Table 4-1 Manufacturer suggested peak flow rates

| Fixture or Appliance | Additional Information | Manufacturer Flow Rate Value Low (GPM) | Manufacturer Flow Rate Value High (GPM) | Maximum Manufacturer Flow Rate Value at 60 PSI (GPM) | |
|---|---|---|---|---|---|
| **Bathroom Fixtures** | | | | | |
| Water Closet | 1.28 GPF Gravity Tank System | 3 | 3 | 3 | |
| Water Closet | 1.0 - 1.28 - 1.6 GPF Pressure-Assisted Flushing Tank (with internal PRV) | 3 | 3 | 3 | |
| Water Closet | 1.1 - 1.28 - 1.7 GPF Flushometer Value System | 25 | 25 | 25 | |
| Urinal | 0.125 GPF Flushometer Value System | 3.5 | 18 | 10 | |
| Urinal | 0.5 - 1.0 GPF Flushometer Value System | 9 | 18 | 10 | |
| Bidet | | 1.5 | 2.2 | 2 | |
| Faucet | Lavatory | 0.5 | 1.5 | 1.5 | * |
| Shower/ Shower Wand | per head (no Bathtub) | 1.5 | 2.5 | 2 | |
| Bathtub 3/4" Supply | With or without showerhead | 13 | 30 | 13 | ** |
| Bathtub 1/2" Supply | With or without showerhead | 5 | 25 | 5.5 | |
| **Kitchen Fixtures - Residential and Commercial** | | | | | |
| Faucet | Kitchen Sink | 1.5 | 2.2 | 2.2 | * |
| Faucet | Pot Filler | 1.5 | 6 | 4 | ** |
| Faucet | Bar Sink | 0.5 | 2.2 | 1.8 | * |
| Faucet | Hand Sink | 0.5 | 10 | 1.8 | * |
| Faucet | Commercial Pre-rinse Spray Valves | 1.28 | 5 | 1.28 | |
| Dishwasher | Residential Dishwasher | 1 | 3 | 1.3 | |
| Dishwasher | Commercial - Conveyor | 2.5 | 2.5 | 2.5 | |
| Dishwasher | Commercial - Single Rack Dishwashers | 1 | 1 | 1 | |
| Dishwasher | Commercial - Double Rack Dishwashers | 1.5 | 1.5 | 1.5 | |
| Dishwasher | Commercial - Under the Counter/ Glass Washer | 0.5 | 0.5 | 0.5 | |
| Dipper Wells | Flow continuously to clean utensils | 0.3 | 1 | 0.5 | |
| Ovens/Steamers | Connected to water supply | 1 | 1 | 1 | |
| **Laundry/Utility Fixtures** | | | | | |
| Clothes Washer | Residential | 3.5 | 3.5 | 3.5 | |
| Clothes Washer | Commercial - Very dependent on fill level and model | 6 | 75 | 12 | ** |
| Faucet | Laundry | 1.8 | 4 | 2 | |
| Faucet | Utility/Mop/Service | 2.2 | 20 | 15 | ** |
| Wash-Down Sprayers | Flow rate dependent on if used with a nozzle. | 6 | 20 | 6 | ** |
| **Specialized Medical Industry Fixtures** | | | | | |
| Bedpan washer | | 1.5 | 4.8 | 1.5 | ** |
| Faucet | Medical Scrub Sink - Wash up Sink | 4.5 | 4.5 | 4.5 | |
| Faucet | Medical Scrub Sink - Infection Prevention Sink | 1.5 | 1.5 | 1.5 | |
| Faucet | Clinical Sink | 1.2 | 2.2 | 1.5 | |
| Faucet | Clinical Sink Flush Valve | 25 | 25 | 25 | |
| Steam Sterilizers | | 1 | 3 | 1 | |
| Dental Unit | | 1 | 2 | 1 | |
| **Miscellaneous Fixtures** | | | | | |
| Ice Maker/soda fountain/ coffee machine | | 0.5 | 10 | 1.75 | ** |
| Drinking Fountain | | 0.75 | 1.5 | 1 | |
| Water Bottle Dispenser | | 0.5 | 1.5 | 1 | |

* Faucet aerators can be used to reduce the water flow rate. They are commonly available with ratings of 0.5, 0.75, and 1.0 GPM.

** There are a wide variety of flow rates for this fixture. Manufacturer specification sheets are recommended.

Information for this table was obtained from commercial and residential suppliers, major manufacturers and EPA WaterSense documentation and specification sheets.

When known, the published fixture values should be used when considering the residual pressure at the fixture outlet. If the actual building plumbing system pressure fluctuates higher or lower than the manufacturer's recommended published flow rate and pressure, there would be a variation in the performance of the fixture. For example, WaterSense-certified faucets and faucet accessories are required to not exceed a flow rate of 1.5 gpm at a pressure of 60 psi at the inlet. Fixture flow rates generally increase with system pressures (see Figures 4-4 and 4-5). Additionally, when calculating the fixture value of a plumbing fixture, it is crucial that the entire fixture is evaluated. For instance, a showerhead or hand shower fitting with a bath trim kit should be assessed on the highest instantaneous demand for the entire fixture. Even if the showerhead is rated for 1.5 gpm at 60 psi, the tub demand could fluctuate significantly. For instance, a standard tub has a flow rate around 5 gpm (see Figure 4-4), while a tub spout with a larger plumbing supply or one that is designed to fill a Roman-style tub could have a flow rate approaching 20 gpm (see Figure 4-5).

There are fixture features (e.g., nozzles and aerators) that can be used to reduce the flow rate and pressure, but it is up to the utility or the water system designer to determine what, if any, flow rate reductions are appropriate when calculating the final peak water demand for the building. Calculation for peak water demand in a nonresidential building is a three-step process, as illustrated in Figure 4-6.

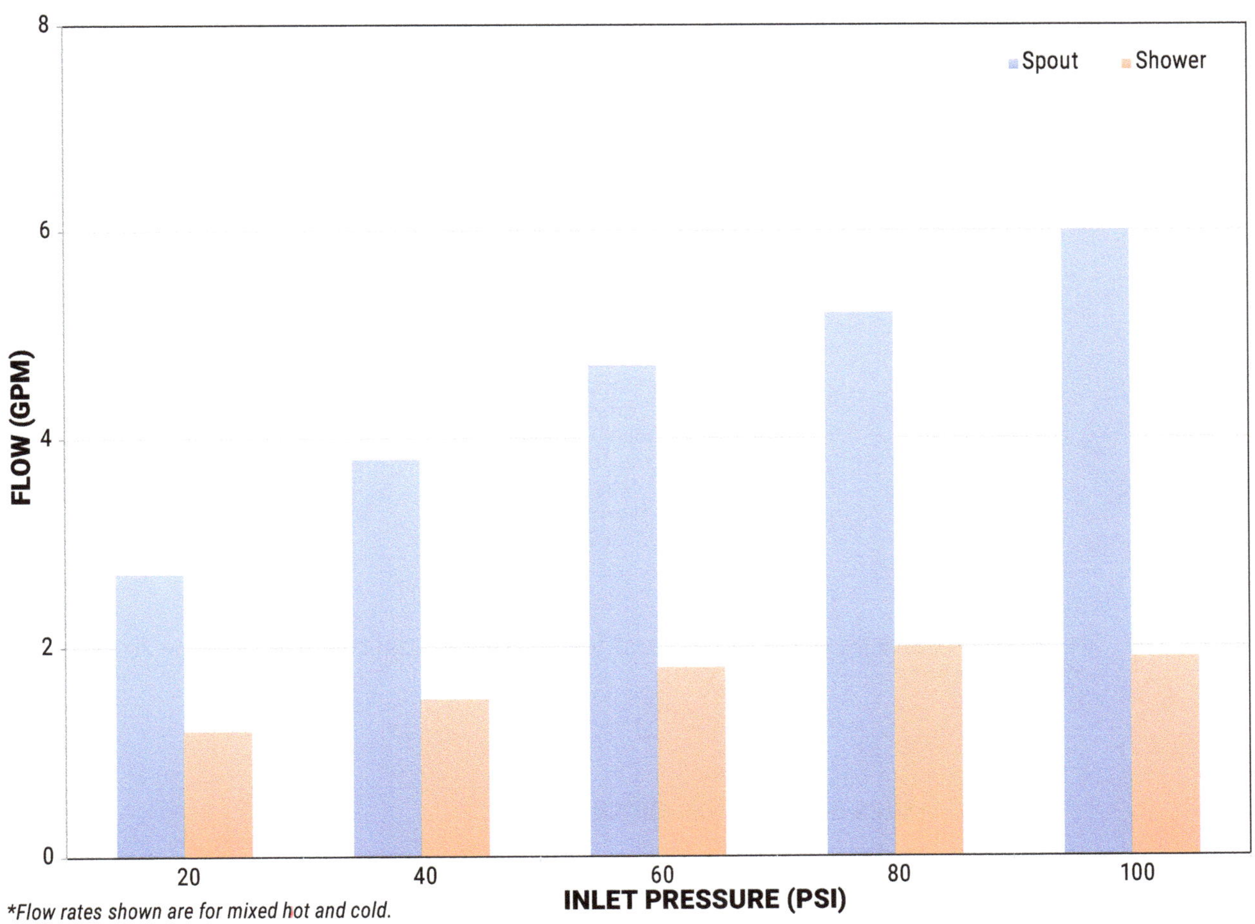


Figure 4-4 Bath/shower trim kit fixture example

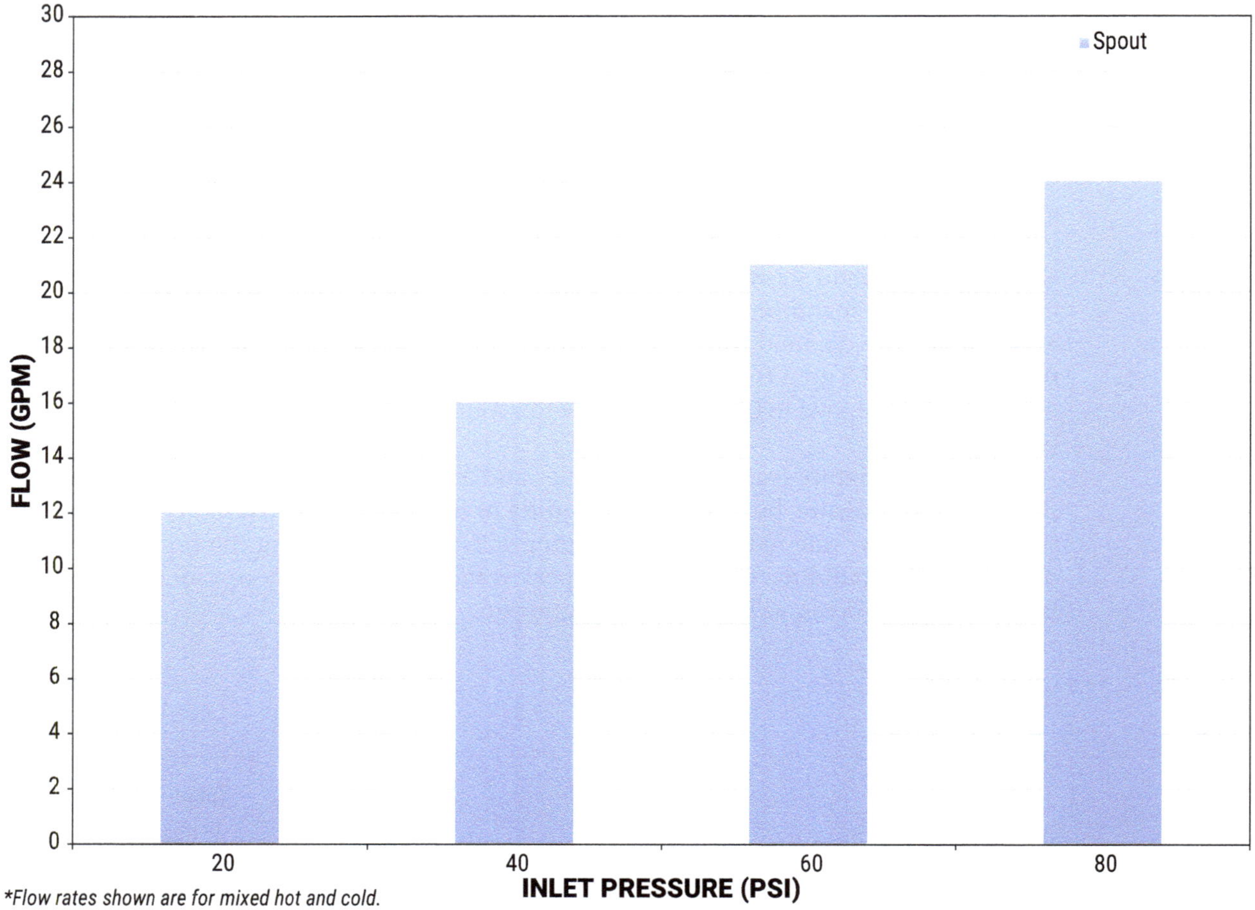


Figure 4-5 Roman tub flow rate (American Standard T186900)

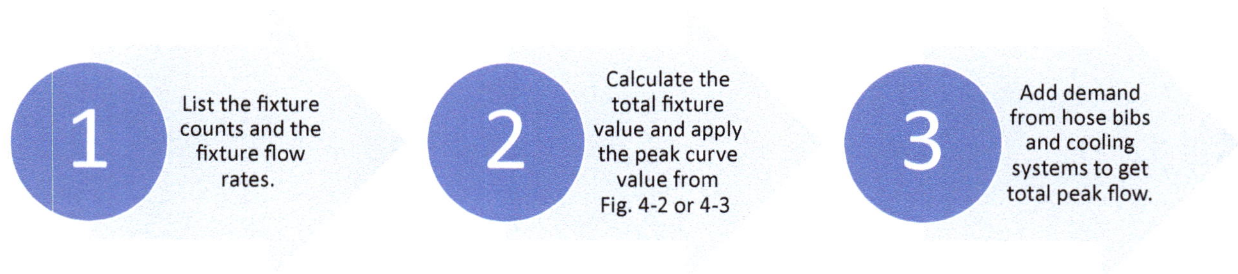


Figure 4-6 Steps to calculate peak water demand for nonresidential buildings via the fixture value approach

## Practical Issues and Legal Ramifications

The most common types of fixtures and uses have been identified in an effort to help the water system designer or the water utility estimate the peak indoor demand from nonresidential customers. While this information is expected to cover the vast majority of cases, it is possible that scenarios will arise where the actual user demand will exceed the estimated peak demand. Hence, it is important that the utility or water system designer thoroughly evaluates the customer's future potential demand, demand pattern, and plumbing supply before finalizing a recommendation. As part of this process, it is imperative that the water utility or designer is aware and knowledgeable about the size of the plumbing supply, the pressure available at the meter, the head loss due to pipe friction, minor losses and elevation changes, and the water velocity that will occur during the peak flow rate.

Depending on the jurisdictional relationship between the utility and the local municipality, the utility may be unable to inspect and verify that the fixtures and appliances installed agree with those that were listed in the building plans during the design review. Additionally, depending on permitting, the utility may not be notified if future changes are made to fixtures or if additional fixtures are added to the building. The utility should evaluate the customer's current and future occupancy and the typical demand profile of similar building types before reaching a decision on the final calculated peak water demand and meter sizing. It remains the responsibility of the water system designer to recognize, identify, and apply the appropriate flow rates needed to generate a reasonable and representative combined fixture value (gpm) for use in the new design charts shown in Figures 4-2 and 4-3.

Finally, it is important to recognize that there are many municipal users, commercial businesses, and specialized industries with particular water needs and uses such as water parks, high-tech manufacturing, bottling plants, processing facilities, and many others that are beyond the scope of this chapter. Such cases need to be carefully considered on an individual basis.

## Example of Peak Demand Calculation for a Nonresidential Building

The following example demonstrates how to use the fixture value approach for a nonresidential building. The 157-room hotel in this example was designed using LEED certification standards and reviewed and built in Denver between 2016 and 2018. The residual pressure at the main was estimated to be approximately 76 psi. The original meter sizing calculation was performed using the fixture value approach from the third edition of M22 (AWWA 2014). The peak demand, including hose bibb flow, was calculated to be 100 gpm. Based on applicable standards at the time, the hotel was sized with a 2-in. meter when submitted for review to Denver Water.

Figure 4-7 shows calculations for the new peak demand at the hotel based on the updated design curves following the three-step procedure described in Figure 4-6. Using the updated curves, the final peak demand is 88 gpm, a decrease of 12 gpm from the original calculation based on the third edition of M22.

This peak flow estimate includes the total continuous demand of five ½-in. hose bibbs. The example in Figure 4-7 uses the maximum manufacturer flow rate value at 60 psi for each fixture type listed in Table 4-1. The lone exception to this is the group of three 65-lb clothes washing machines, which were sized using the manufacturer specification sheets and information provided by the designer. As mentioned in the footnotes for Table 4-1, if the actual flow rate for a specific plumbing fixture is known (and if it can be verified that the fixture was installed), then the actual flow rate should be used instead of flow values listed in Table 4-1.

In this example, the hotel has five hose bibbs available for washing down the property and for other outdoor uses. The bibbs are typically placed around the perimeter of the building to avoid long hose pulls. Since loads from hose bibbs are considered continuous demand, this outdoor demand must be added to the peak indoor demand estimated from the fixture value design curves. In this case, the five hose bibbs at the hotel produce 25 gpm of outdoor water use and, when added to the indoor peak of 63 gpm, contribute to an overall total building peak demand of 88 gpm.

**Step 1: List the fixture counts and the fixture flow rates**

| Fixture or Appliance | Additional Information | Number of Fixtures | Manufacturer Flow Rate (GPM) | Total Fixture Value (GPM) ★ |
|---|---|---|---|---|
| Water Closet / Toilet | Pressure-Assisted Flushing Tank | 159 | 3 | 477 |
| Water Closet / Toilet | Flushometer Value System | 8 | 25 | 200 |
| Shower/ Shower Wand | per head (no Bathtub) | 158 | 2 | 316 |
| Faucet | Lavatory | 169 | 1.5 | 253.5 |
| Faucet | Kitchen Sink | 164 | 2.2 | 360.8 |
| Faucet | Utility/Mop/Service | 3 | 15 | 45 |
| Dishwasher | Commercial - Single Rack Dishwashers | 2 | 1 | 2 |
| Dishwasher | Residential | 157 | 1.3 | 204.1 |
| Clothes Washer | Commercial - 13lb | 3 | 12 | 36 |
| Clothes Washer | Commercial - 65lb | 3 | 17 | 51 ★★ |
| Drinking Fountain | | 1 | 1 | 1 |

**Step 2: Calculate the total fixture value and apply to the peak flow curve**

| | |
|---|---|
| Total Fixture Value: | 1,946 |
| Demand (GPM) from Fig 4-2 | 63 |

**Step 3: Add demand for hose bibbs and cooling systems to get total peak flow**

| Fixture or Appliance | Additional Information | Number of Fixtures | Flow Rate (GPM) | Total Fixture Value (GPM) |
|---|---|---|---|---|
| Hose Bibb 1/2" | | 5 | 5 | 25 ★★★ |
| | | | | 88 |

★ The manufacturer flow rate is based on Table 4-1, Maximum Manufacturer Flow Rate Value at 60 PSI

★★ The 65-lb washing machine, based on the manufacturer's published rates, can accommodate any fill rate between 4 - 85 GPM. The manufacturer's published rates are based on 2 - ¾" connections. The design engineer created a 1.5" manifold system to limit the flow to 17 GPM per machine.

★★★ Based on the building use, the water design engineer or the utility may want to consider a reduction in the number of hose bibbs counted towards the total probable demand. The hose bibbs, like in the case of a hotel, may be used for washing down the property and are typically placed strategically around the building to avoid long hose pulls.

**Figure 4-7 Sample 157-room hotel peak water demand calculation using three-step process**

Since it is unlikely that all five hose bibbs will be flowing simultaneously, the water system designer or the utility may want to consider a reduction in the number of hose bibbs counted toward the peak demand. This becomes particularly important as the size of the property increases. A large number of hose bibbs in the design could lead to an artificially high estimate of the peak water demand. This, in turn, can inadvertently increase the meter size, which could cause the loss of a meter's low-flow accuracy—resulting in unregistered and unbilled water and a loss of revenue, as discussed in Chapter 3.

When the new building opened for business, Denver Water (with the cooperation of hotel management and the water system designer) monitored water use for 46 days at 10-s intervals with a digital input data logger linked to a magnetic pulse sensor on the 2-in. meter. Figure 4-8 shows daily peak instantaneous flow and the daily hotel occupancy rate from Nov. 1, 2019, through Dec. 16, 2019. Since the hotel management was interested in identifying morning peak (12 a.m.–8 a.m.), laundry hours (8 a.m.–4:30 p.m.), and evening peak periods (4:30 p.m.–12 a.m.), the graph shows peak flows during those time intervals and highlights when the daily peak was reached. For this hotel, 74 percent of the daily

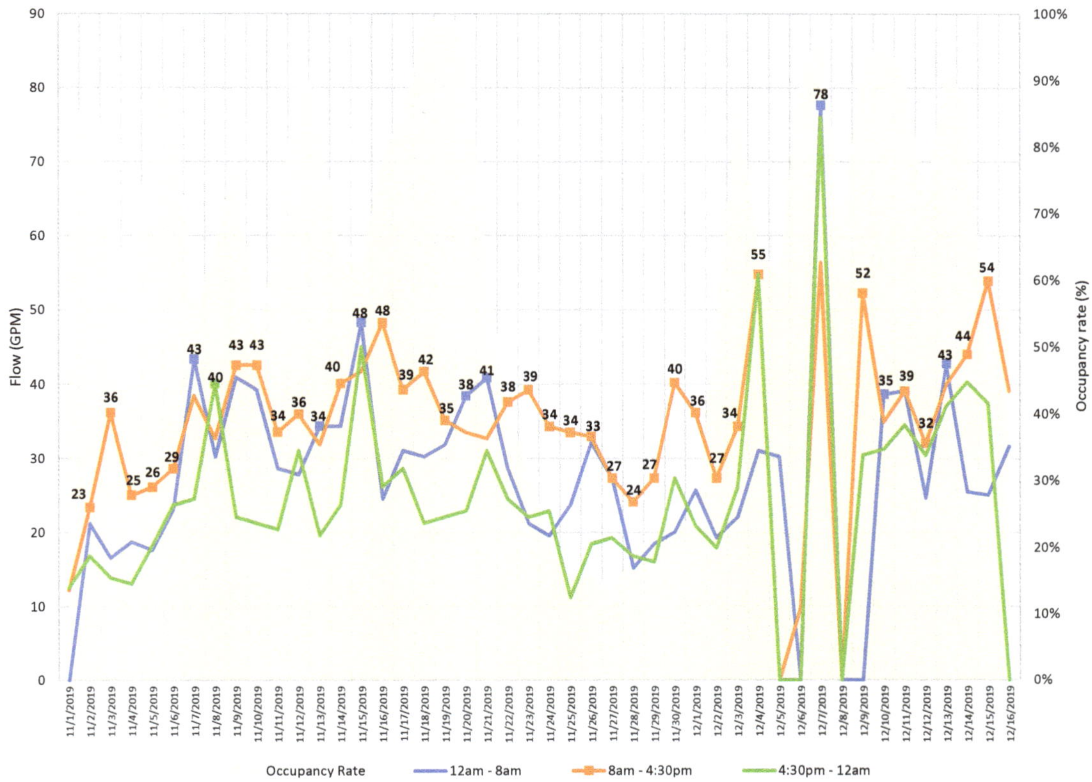


**Figure 4-8 Hotel nonresidential example maximum flow during specific daily time intervals from Nov. 1, 2019 through Dec. 16, 2019**

peak flow rates occurred during the laundry hours (8 a.m.–4:30 p.m.). There was a single water use spike of 78 gpm on Dec. 7. After coordinating with the hotel staff, Denver Water learned the spike was because of a boiler issue that forced the hotel staff to flush hot water lines throughout the building. Coincidentally, during the same time period (Dec. 5, 6, 8, and 9), the loggers lost data. This malfunction was likely because of cold temperature conditions.

Figure 4-9 shows the relative frequency (left side) and cumulative frequency (right side) plots of the hotel flow data sorted into 1 gpm bins. These data represent all the flows measured during the 46-day period from Nov. 1 to Dec. 16, 2019, at the 2-in. meter in the 157-room hotel (excluding the 4-day malfunction episode in early December). During this period, hotel occupancy ranged from 15 percent to 100 percent with an average occupancy of about 60 percent, as shown in Figure 4-8.

At the low end, no flow (0 gpm) was registered nearly 8 percent of the time. The average of all recorded data is 11.1 gpm, and the 50th percentile (median) is 9.7 gpm. At the high end, the 90th percentile of the hotel flows is 24 gpm, while the 99th percentile of all measured flows is 35 gpm.

Considering only the daily peaks (43 values in Figure 4-8), the average and the median of the daily peak flow are both close to 38 gpm. The 99th percentile of the daily peak is 56 gpm (compared to the estimate of 88 gpm in Figure 4-7). All observed demands remained well below the meter's Safe Maximum Operating Capacity (SMOC) of 170 gpm. All things considered (meter size, daily peaks, nonirrigation season), the new design curves in Figures 4-2 and 4-3 produced an estimate of the peak demand at this hotel with a reasonable degree of conservatism.

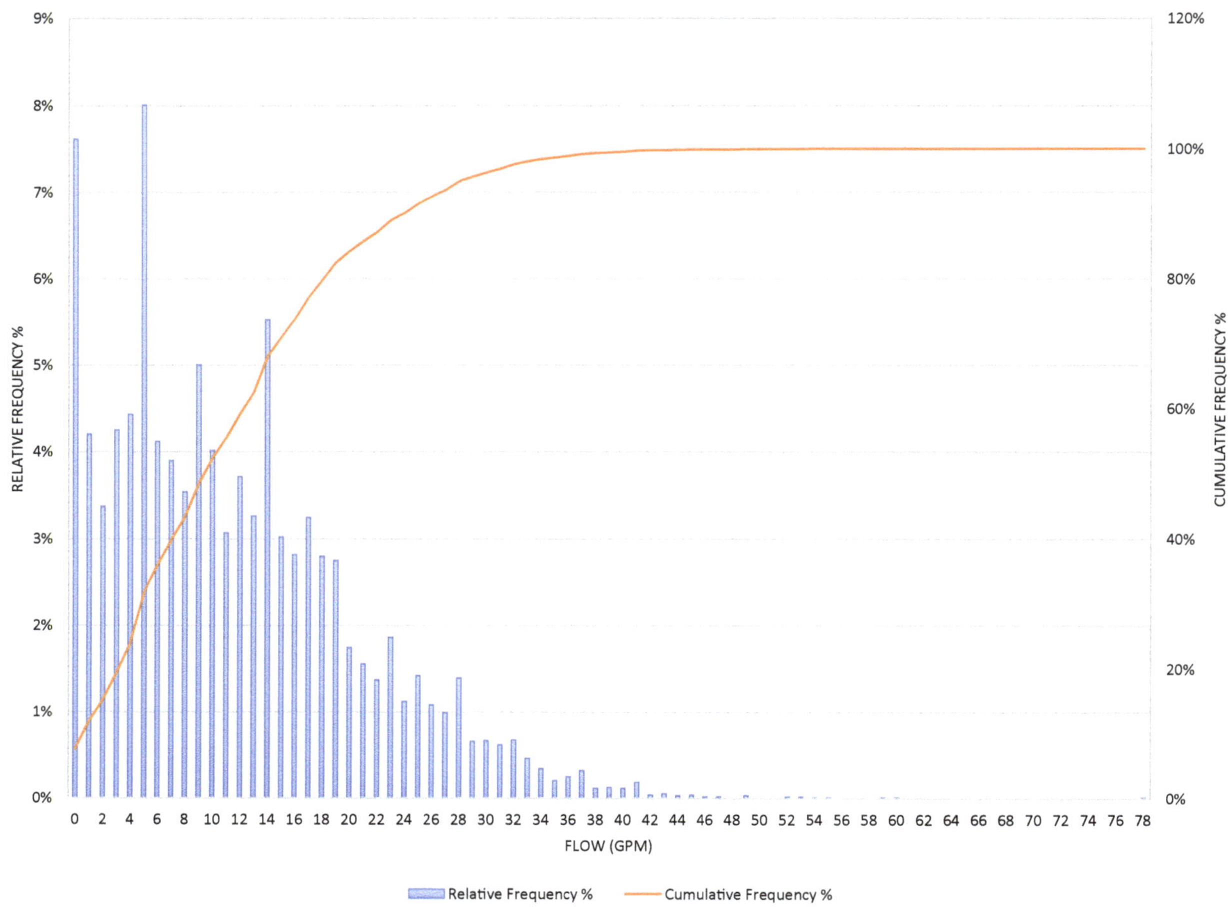


**Figure 4-9 Histograms: Hotel high-resolution flow data (Nov. 1, 2019 through Dec. 16, 2019)**

## Estimating Peak Demand for Residential Buildings

As illustrated in the previous section, peak water demands in nonresidential buildings are estimated using an updated version of the traditional M22 design graphs based on total fixture values. Nonresidential buildings may include flush-valve fixtures and flush-tank fixtures. Residential buildings are assumed to have flush-tank fixtures. In contrast, the new procedure for estimating peak water demands in residential buildings uses a probabilistic framework, similar in concept to Hunter's method, but without the need for fixture units or fixture values. At present, the new peak flow method for residential buildings considers only flush-tank fixtures; flush-valve fixtures are not (yet) allowed.

The new probabilistic approach is built around three algorithms developed to accommodate small, medium, and large residential buildings. Small buildings are those with 30 or fewer indoor fixtures. Large residential buildings are those with several hundred indoor fixtures (the threshold is not exact in terms of the fixture count). Medium-size buildings are wedged between the small and large categories. The three algorithms for estimating peak water use are programmed into a convenient web-based design aid called the Water Demand Calculator (WDC). This section gives a brief review of Hunter's key insight; introduces the three algorithms in the WDC; and presents some examples of the WDC applied to residential buildings with water-conserving fixtures in the United States and Australia.

## Hunter's Method

Roy Hunter (1940) demonstrated that the number of busy water fixtures in a building can be described with the binomial probability distribution. Given a group of $n$ identical fixtures each with probability $p$ of being used, Hunter showed the probability of having exactly $x$ fixtures operating simultaneously out of $n$ total fixtures has a binomial mass function:

$$\Pr\left[x\ busy\ fixtures \middle| n, p\right] = \binom{n}{x} (p)^{x} (1-p)^{n-x} \qquad x = 0,1,\ldots,n \qquad \text{(Eq. 4-1)}$$

Most buildings have an assortment of fixtures. Each fixture group has their own unique values for $n$ and $p$ and, hence, has their own distinct version of Eq. 4-1. Hunter used the 99th percentile from Eq. 4-1 as the design standard for estimating peak demand. He recognized, however, that when estimating the load on the building, it was not legitimate to simply add the 99th percentile from each fixture group. In a clever move, Hunter introduced *fixture units* to merge the 99th percentile curves for each fixture group into a single design curve, giving the 99th percentile of peak demand at a building based on the total fixture units. The final result, called Hunter's curve, is the theoretical basis for plumbing codes around the world (IAPMO 2018a,b). While scientifically sound, Hunter's Curve is calibrated for conditions in 1940 and, as explained in the beginning of this chapter, it does not reflect water use behavior in new buildings with efficient fixtures.

## Wistort's Method (WDC Algorithm for Large Residential Buildings)

Robert Wistort (1995) proposed using the normal approximation for the binomial distribution to estimate directly peak loads on a plumbing system. Similar to Hunter's approach, the number of busy fixtures $x$ is considered to be a random variable with a binomial distribution having a mean $E[x] = np$ and variance $Var[x] = np(1-p)$. From the normal approximation, the estimate of the 99th percentile of the demand in a building with $K$ different fixture groups is:

$$Q_{0.99} = \sum_{k=1}^{K} n_k p_k q_k + (z_{0.99}) \sqrt{\sum_{k=1}^{K} n_k p_k (1-p_k) q_k^2} \qquad \text{(Eq. 4-2)}$$

In this expression, $n_k$ is the total number of fixtures belonging to fixture type $k$; $p_k$ is the probability that a single fixture in fixture type $k$ is operating during the peak period; $q_k$ is the flow rate at the busy fixture type $k$; and $Z_{0.99}$ is the 99th percentile of the standard normal distribution. Wistort figured out how to add the demands of each type of fixture (e.g., the demand from all the kitchen sinks, the demand from all the bathroom sinks, the demand from all the showers) directly by approximating the binomial distribution as a normal distribution. The chief advantage of Wistort's method is that it avoids the need for fixture units and is readily extended to other types of fixtures, provided values for $p$ and $q$ are available.

Wistort's method is most suitable when the dimensionless *Hunter Number*, H, equals or exceeds five:

$$\mathrm{H} = \sum_{k=1}^{K} n_k p_k \geq 5 \qquad \text{(Eq. 4-3)}$$

H represents the average number of simultaneous busy fixtures in the building during the peak period (Buchberger et al. 2017). In the context of residential plumbing, fixture $p_k$ values tend to be small (average $p \approx 0.02$) even during peak periods, so the total fixture count must be large ($\Sigma n \geq 250$) to satisfy Eq. 4-3. Accordingly, Wistort's method can be used for estimating peak demands in large buildings with many residential units, but it is not appropriate for small residential buildings.

## Modified Wistort Method (WDC Algorithm for Medium-Size Residential Buildings)

In a building with $K$ different and independent fixture groups, the probability, $P_0$, that all fixtures in the building are idle (i.e., zero demand at the meter) during the period of peak water use is:

$$P_0 = \prod_{k=1}^{K} (1 - p_k)^{n_k} \approx \exp(-\mathrm{H}) \qquad \text{(Eq. 4-4)}$$

The result in Eq. 4-4 can be interpreted as the fraction of time that the building meter registers no water use (stagnation) during the peak period. For instance, at a large residential building (e.g., 42 two-bedroom apartments with H = 7), Eq. 4-4 gives $P_0$ <0.1 percent—indicating there is virtually no chance of experiencing peak-period stagnation. In contrast, at an intermediate-size apartment building (e.g., 12 two-bedroom units with H = 2), nearly 15 percent of the peak hour will register zero demand at the building meter.

Premise plumbing systems are not designed for zero flow. The possibility of stagnation during the period of peak demand should not influence the size of the plumbing system. To circumvent the zero-flow issue, a *zero-truncated binomial distribution* (ZTBD) is used to describe the conditional probability distribution of busy fixtures in medium-size residential buildings. Assuming that a normal approximation describes the upper tail of the ZTBD, the expression for the 99th percentile of the demand in a residential building with $K$ different fixture groups is:

$$Q_{0.99} = \frac{1}{1-P_0}\left[\sum_{k=1}^{K} n_k p_k q_k + \left[(1+P_0) z_{0.99}\right]\sqrt{\left[(1-P_0)\sum_{k=1}^{K} n_k p_k (1-p_k) q_k^2\right] - P_0\left(\sum_{k=1}^{K} n_k p_k q_k\right)^2}\,\right] \qquad \text{(Eq. 4-5)}$$

When $\mathrm{H} > 5$, $P_0 \to 0$ and Eq. 4-5 reduces to Eq. 4-2. For this reason, the ZTBD approach is called the "Modified Wistort Method" or MWM (Buchberger et al. 2017). In practice, the transition from medium-size buildings (Eq. 4-5) to large buildings (Eq. 4-2) requires at least 200–300 indoor fixtures. Results of Monte Carlo computer runs to simulate indoor residential water use indicate that the MWM provides reasonable estimates of peak water demand over a wide range of household fixture counts. For $n = 1$ and $k = 1$ (i.e., last fixture at the end of the line in the building, furthest from the water meter), Eq. 4-5 simplifies to $Q_{0.99} = q$. In other words, the design flow is simply the nominal demand of the final fixture. In the WDC, use of the MWM approximation is restricted to medium-size residential buildings. For small residential buildings ($n < 30$ fixtures), an alternate but exact computational procedure is used to estimate the 99th percentile of the peak demand.

## Exhaustive Enumeration (WDC Algorithm for Small Residential Buildings)

Hunter and Wistort focused on the 99th percentile of the demand expected during the peak period in a building. With today's computational tools, it is possible to numerically generate the entire probability distribution of water demands at any point and any time in a building. "Exhaustive Enumeration" involves identifying and ranking all possible demand events for a given premise plumbing configuration. While simple in principle, exhaustive enumeration is not always practical. Because of combinatorial explosion, the size of the problem grows geometrically. For example, a typical single-family home with 12 independent fixtures will generate $2^{12} = 4{,}096$ demand outcomes. So while enumeration can give an exact solution and help to visualize the discrete probability distribution of residential fixture demands, it is restricted to scenarios where the total fixture count $n$ is manageable. In the current version of WDC, exhaustive enumeration is used when the fixture count is 30 or less (small residential buildings only). Additional information including detailed examples featuring exhaustive enumeration can be found in Buchberger et al. (2012), Omaghomi and Buchberger (2014), and Buchberger et al. (2017).

## Method Summary for Residential Buildings

The problem of implementing an analytical framework to estimate peak water demands in buildings is challenging because the solution strategy changes with the size of the building. On a large scale [$n > 200$ fixtures, $H > 5$], individual fixtures do not appreciably affect the performance of the premise plumbing system. In such cases, the Wistort method (Eq. 4-2) based on well-established continuous probability distributions can be applied to estimate peak building demands. On a small scale [$n < 30$ fixtures, $H < 0.5$], individual fixtures can exert a significant impact on the behavior of the premise plumbing system. For this case, exhaustive enumeration is needed to account for the effect of discrete fixture operation on the magnitude of the peak flow in a small premise plumbing system. The modified Wistort method (Eq. 4-5) bridges the gap between the exact but cumbersome discrete math algorithm for small residential buildings and the approximate but expedient continuous math algorithm for large residential buildings.

Table 4-2 summarizes the three algorithms programmed into the WDC for estimating peak demands in residential buildings. The applicability of each algorithm corresponds to a color-shaded region (green, yellow, red) in the tri-color $n$-$p$ plane of Figure 4-10. On the y-axis, there is an order-of-magnitude variation in the probability of fixture use, ranging from $p = 0.005$ (dishwashers) to $p = 0.055$ (washing machines and tub/shower combinations). These fixture p-values were estimated from water-use data collected from over 1,000 single family homes in 63 cities across the United States (Omaghomi et al. 2020). The color-shaded regions in Figure 4-10 are dissected by diagonal lines that represent a constant Hunter Number. The magnitude of H increases from left to right in Figure 4-10 as the number of fixtures (and size of the building) grows.

**Table 4-2** Methods to estimate peak demands in residential buildings with efficient fixtures

| Region in Figure 4-10 | | Building Size | Algorithm in Water Demand Calculator | Boundary Criteria | Probability of Peak Period Stagnation, $P_0$ |
|---|---|---|---|---|---|
| Green | ● | Small | Exhaustive Enumeration [ExEn] | $n < 30$ | Very High |
| Yellow | ● | Medium | Modified Wistort Method [MWM] | $n > 30$<br>$H < 5$ | Moderate |
| Red | ● | Large | Wistort Method [WM] | $H > 5$ | Very Low |

*Source:* Courtesy of Dr. Steven Buchberger

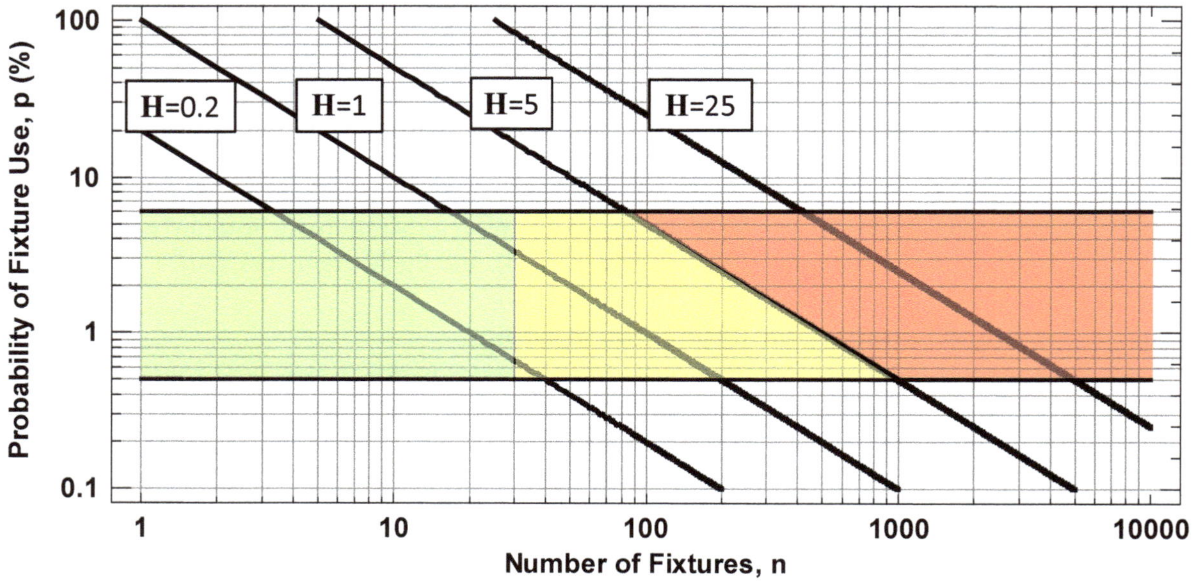


*Source:* Courtesy of Dr. Steven Buchberger

**Figure 4-10** **Water Demand Calculator uses three algorithms to estimate peak flow in residential buildings. ExEn is used in the green region, MWM in yellow, and WM in red, corresponding to small, medium, and large buildings, respectively. Diagonal lines are constant values of the dimensionless Hunter Number, H, representing the average number of simultaneous busy fixtures during the period of peak demand.**

## Water Demand Calculator (WDC)

The three approaches for estimating residential peak water demand can be cumbersome to use, even when the designer knows all the required input parameters (i.e., n, p, q) for each fixture group. As noted earlier, the WDC is the recommended method for estimating peak water demand in residential buildings. The WDC is incorporated into the 2018 Uniform Plumbing Code and is available free as a downloadable Microsoft Office Excel spreadsheet at the IAPMO website: https://www.iapmo.org/water-demand-calculator/.

A screenshot of the template for WDC v2.1 is shown in Figure 4-11. The central core is organized in a spreadsheet format with white and blue-shaded cells. The Probability of Use (p) and Maximum Recommended Fixture Flow Rate (q) values in the blue-shaded cells are derived from a national survey of indoor water use at hundreds of homes with efficient fixtures. These values cannot be changed. The white cells accept input from the designer. Across the top of the WDC template are several input fields to specify a project name and indicate the type of building (single family or multifamily). The multifamily option opens two additional input fields to specify: the total number of apartments (or condominiums, flats, etc.) in the building and the total number of apartments that are to be used in the peak flow computation. These two counts are not necessarily the same. For instance, in a large residential building with 150 apartments spread over three separate wings, the designer may be interested in the peak demand originating from the 50 apartments in the west wing only.

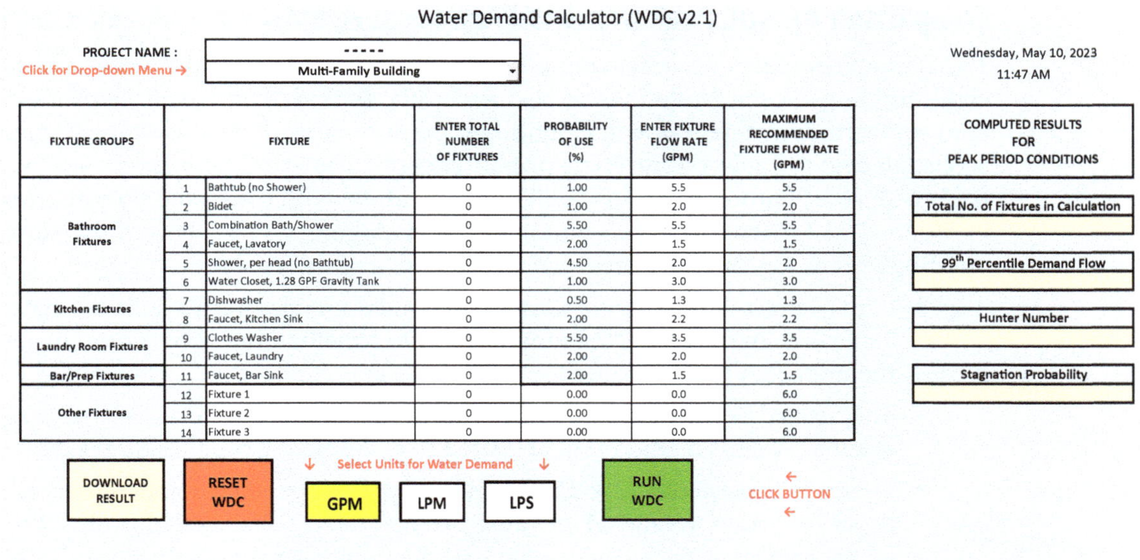


| FIXTURE GROUPS | | FIXTURE | ENTER TOTAL NUMBER OF FIXTURES | PROBABILITY OF USE (%) | ENTER FIXTURE FLOW RATE (GPM) | MAXIMUM RECOMMENDED FIXTURE FLOW RATE (GPM) |
|---|---|---|---|---|---|---|
| Bathroom Fixtures | 1 | Bathtub (no Shower) | 0 | 1.00 | 5.5 | 5.5 |
| | 2 | Bidet | 0 | 1.00 | 2.0 | 2.0 |
| | 3 | Combination Bath/Shower | 0 | 5.50 | 5.5 | 5.5 |
| | 4 | Faucet, Lavatory | 0 | 2.00 | 1.5 | 1.5 |
| | 5 | Shower, per head (no Bathtub) | 0 | 4.50 | 2.0 | 2.0 |
| | 6 | Water Closet, 1.28 GPF Gravity Tank | 0 | 1.00 | 3.0 | 3.0 |
| Kitchen Fixtures | 7 | Dishwasher | 0 | 0.50 | 1.3 | 1.3 |
| | 8 | Faucet, Kitchen Sink | 0 | 2.00 | 2.2 | 2.2 |
| Laundry Room Fixtures | 9 | Clothes Washer | 0 | 5.50 | 3.5 | 3.5 |
| | 10 | Faucet, Laundry | 0 | 2.00 | 2.0 | 2.0 |
| Bar/Prep Fixtures | 11 | Faucet, Bar Sink | 0 | 2.00 | 1.5 | 1.5 |
| Other Fixtures | 12 | Fixture 1 | 0 | 0.00 | 0.0 | 6.0 |
| | 13 | Fixture 2 | 0 | 0.00 | 0.0 | 6.0 |
| | 14 | Fixture 3 | 0 | 0.00 | 0.0 | 6.0 |

*Source:* Water Demand Calculator v2.1

**Figure 4-11 Screenshot of the input/output template for the Water Demand Calculator v2.1**

Across the bottom of the WDC template are several multicolored executable click buttons that allow the user to download results, reset (clear) the template, select the system of flow units, and run the WDC program. Stacked along the right side of the WDC template are four boxes that display computed output values for peak flow conditions in the building: the total fixture count, the estimated 99th percentile of the peak period demand, the Hunter Number (H), and the estimated probability of stagnation (i.e., fraction of time during peak period when there is zero demand, as predicted by Eq. 4-4).

In the central spreadsheet part of the WDC template, there are 11 indoor residential fixtures, grouped by room. Similar to the example for the nonresidential building earlier in this chapter, the user must provide the total count for each fixture. There are three open fields (white cells in lines 12-14) that allow the user to specify other indoor fixtures, along with their corresponding p-value and q-value. The current version of the WDC has upper limits on the magnitude of the user-specified p and q values. The p-values cannot exceed 6 percent for any indoor fixture; the maximum allowable q-values are listed in the rightmost column of the input spreadsheet. The WDC will issue an error message if these limits are exceeded.

When all necessary input fields in Figure 4-11 are complete, click the RUN WDC button. The WDC will select the appropriate algorithm, per Table 4-2, and provide an estimate of the 99th percentile flow expected during the peak period. Outdoor water use from hose bibbs and irrigation systems is not computed with WDC. Similar to the nonresidential case, outdoor residential water use, if concurrent, must be added to the estimated 99th percentile of the indoor demand found from running the WDC.

In addition to the input/output template tab shown in Figure 4-11, the Excel file for WDC v2.1 includes two other tabs. There is a Title Page tab with some basic background information and a Key tab with helpful details about the features on the WDC input/output template.

## Comparison of WDC Peak Demand Predictions Against Field Measurements

This section compares peak demand estimates from the WDC against flows measured at eight buildings in the United States and Australia. Peak demands and other interesting relevant features at new residential buildings are summarized in Table 4-3 for Denver and Scottsdale and in Table 4-4 for Sydney and Canberra. The WDC templates showing input parameters (i.e., fixture counts and building size) along with computed output are shown in Figure 4-12 for two of the US buildings and in Figure 4-13 for two of the Australian buildings. Some key observations from these tables and figures are:

- Both Denver locations are classified as medium-size residential buildings (see Table 4-2). Accordingly, the modified Wistort method (Eq. 4-5) is used to generate the predicted peak demand.
- The Scottsdale site and all four buildings in Australia are classified as large; the original Wistort method (Eq. 2) is used to compute the predicted peak demand.
- There is consistently good agreement between predicted and observed peak flows (see rows 14 and 15 in Table 4-3 and Table 4-4). Normal probability plots of measured peak period residential flows in Denver, Sydney, and Canberra are shown in Figures 4-14 and 4-15. Both figures provide additional compelling evidence that the WDC provides a reasonable estimate of the 99th percentile of the peak period indoor water demand.
- There is good agreement between predicted and observed peak hour stagnation (compare rows 11 and 12 in Table 4-3 and Table 4-4). The large buildings have virtually no chance of stagnation during the peak period. The occurrence of peak period stagnation at Denver Building 119 is predicted and evident in Figure 4-14(a) and Figure 4-14(b).
- Row 10 in Tables 4-3 and 4-4 shows "peak period probability" (3P) as ratio of Hunter Number to Total Fixture Count, H:Sn. This parameter tends to hover near 1 percent with a slight inverse relationship with building size. There are two interesting interpretations: First, on average, about 1 percent of all fixtures in a residential building are in simultaneous use during the peak period. Second, the building-wide probability that any individual fixture will be in use at any instant during the peak period is about 1 percent. Conversely, even during the peak period, the vast majority of water fixtures in a residential building are idle at any instant.
- The fixture p-values decline slightly as building size increases. For instance, compare the probability of use for a shower at a medium-size Denver Building ($p = 0.0140$) against the large Canberra Building ($p = 0.0065$). This occurs because residential units operate independently. While there is a prevailing diurnal pattern, water-use activities inside residential units are not synchronized. For further insight on this phenomenon, see Omaghomi et al. (2020).

Figure 4-14(a) shows 21 h of peak flow formed by concatenating measurements recorded every 10 s during the peak hour of each day over a three-week period. This new time series is assumed to be a stationary realization of peak-hour demands. Note there are many instances of zero flow. The vertical spike at zero demand gives the frequency of stagnation.

Figure 4-15(a) shows normal probability plots of daily peak demands measured at two large multifamily buildings in Sydney ($n_1 = 214$ days and $n_2 = 150$ days); Figure 4-15(b) shows similar data for two large multifamily buildings in Canberra ($n_1 = 91$ days and $n_2 = 57$ days). The results show that the WDC provides a reasonable estimate of the peak demand in residential buildings fitted with water-conserving fixtures.

**Table 4-3** Peak indoor demands and other features of three residential buildings fitted with water-conserving fixtures at two cities in the United States [*]

| Row | Feature | | | |
|---|---|---|---|---|
| 1 | Location of Building | Denver 114 | Denver 119 | Scottsdale 2 |
| 2 | Monitor Start Date<br>Monitor End Date | 07/06/2017<br>07/21/2017 | 06/16/2017<br>07/06/2017 | 11/17/2017<br>11/30/2017 |
| 3 | Length of Record | 16 days | 21 days | 14 days |
| 4 | Number of Peak Hour Observations | 5,760 | 7,560 | 5,040 |
| 5 | Number of Apartments | 23 | 26 | 224 |
| 6 | Occupancy Rate for Building | 95% | 95% | 95% |
| 7 | Number of Fixtures, Sn | 142 | 161 | 1,658 |
| 8 | Total Fixture Values (gpm) | 415 | 471 | 4,674 |
| 9 | Peak Period Hunter Number, H | 1.74 | 1.92 | 13 |
| 10 | Peak Period Probability 3P = H:Sn (%) | 1.23 | 1.19 | 0.78 |
| 11 | Peak Period Stagnation Frequency Predicted | 17% | 14% | <0.1% |
| 12 | Peak Period Stagnation Frequency Observed | 22% | 12% | 0.2% |
| 13 | Average Daily Peak Flow Observed (gpm) | 11.0 | 11.4 | 42.4 |
| 14 | Maximum Daily Peak Flow Observed (gpm) | **17.2** | **15.1** | **60.2** |
| 15 | WDC version 2.1 Predicted 99th Percentile Demand Flow (gpm) | **19.8** | **20.6** | **70.8** |
| 16 | DIN 1988-300 Predicted 99th Percentile Demand Flow (gpm)[†] | 28.7 | 29.8 | 54.2 |
| 17 | Computational Algorithm, Table 4-2 | MWM | MWM | WM |

**Notes:**

* Demand data are from WRF Project 4689 (Mayer et al. 2020) with 10-s logging interval.

† DIN 1988-300. 2012. Deutsches Institut für Normung (German National Standard 1988-300). The values in row 16 represent the peak water demand in gpm estimated by the German National Standard and are computed with the following expression:

$$Q = 13.88\left[Total\ Fixture\ Values\ in\ Row\ 8\right]^{0.19} - 14.90 \qquad (gpm)$$

**Table 4-4** Peak indoor demands and other features of four residential buildings fitted with water-conserving fixtures at two cities in Australia [*]

| Row | Feature | | | | |
|---|---|---|---|---|---|
| 1 | Location of Building | Sydney 1 | Sydney 2 | Canberra 1 | Canberra 2[†] |
| 2 | Monitor Start Date<br>Monitor End Date | 08/20/2019<br>03/14/2020 | 10/17/2019<br>03/14/2020 | 12/13/2019<br>03/14/2020 | 01/17/2020<br>03/14/2020 |
| 3 | Length of Record | 207 days | 150 days | 92 days | 57 days |
| 4 | Number of Peak Hour Observations | 12,420 | 9,000 | 5,520 | 3,420 |
| 5 | Number of Apartments | 143 | 123 | 330 | 115 |
| 6 | Occupancy Rate for Building | 90% | 90% | 95% | unknown |
| 7 | Number of Fixtures | 1,291 | 1,223 | 2,912 | 859 |
| 8 | Total Fixture Values (gpm) | 2,326 | 2,202 | 5,240 | 1,633 |
| 9 | Peak Period Hunter Number, H | 11 | 10 | 21 | 8.3 |
| 10 | Peak Period Probability 3P = H:Sn (%) | 0.85 | 0.86 | 0.72 | 0.97 |
| 11 | Peak Period Stagnation Frequency Predicted | <0.1% | <0.1% | <0.1% | <0.1% |
| 12 | Peak Period Stagnation Frequency Observed | <0.1% | <0.1% | <0.1% | <0.1% |
| 13 | Average Daily Peak Flow Observed (gpm) | 23.8 | 29.8 | 52.3 | 26.2 |
| 14 | Maximum Daily Peak Flow Observed (gpm) | <u>36.5</u> | <u>49.1</u> | <u>72.9</u> | <u>37.3</u> |
| 15 | WDC version 2.1 Predicted 99th Percentile Demand Flow (gpm) | <u>42.2</u> | <u>41.2</u> | <u>71.5</u> | <u>35.1</u> |
| 16 | DIN 1988-300 Predicted 99th Percentile Demand Flow (gpm)[††] | 45.6 | 45.0 | 55.7 | 41.7 |
| 17 | Computational Algorithm, Table 4-2 | WM | WM | WM | WM |

**Notes:**

* Demand data are from Hydraulic Consultants Association of Australasia https://www.hcaa.org.au/

† Canberra Building 2 is mixed use (820 residential fixtures and 39 commercial fixtures serving 15 businesses, e.g., restaurants, bistros, retailers, jeweler, hairdresser, dry cleaner, optometrist).

†† DIN 1988-300. 2012. Deutsches Institut für Normung (German National Standard 1988-300). The values in row 16 represent the peak water demand in gpm estimated by the German National Standard and are computed with the following expression:

$$Q = 13.88\left[\textit{Total Fixture Values in Row } 8\right]^{0.19} - 14.90 \qquad (gpm)$$

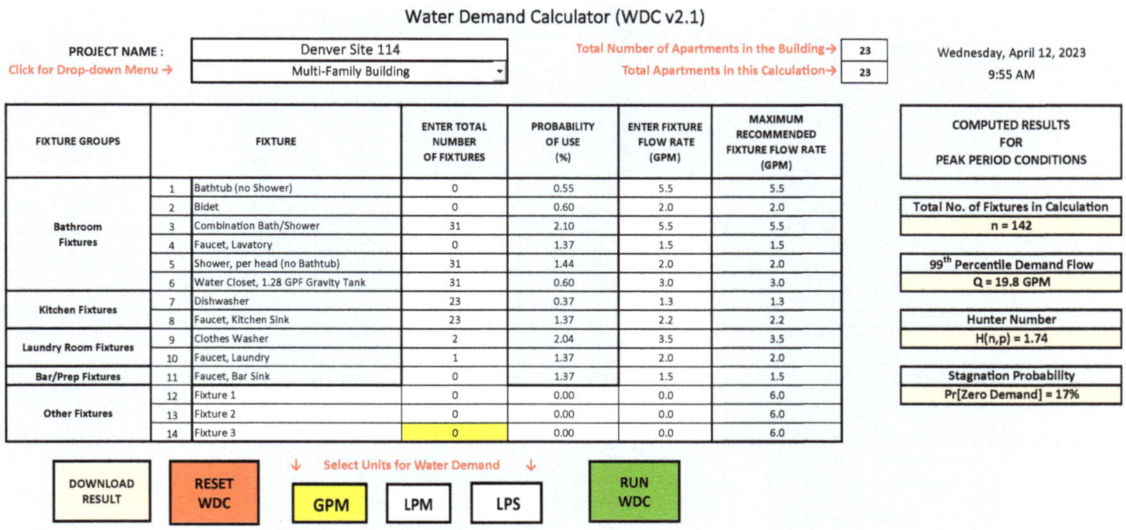

Water Demand Calculator (WDC v2.1)

PROJECT NAME : Denver Site 114

Click for Drop-down Menu → Multi-Family Building

Total Number of Apartments in the Building→ 23

Total Apartments in this Calculation→ 23

Wednesday, April 12, 2023

9:55 AM

| FIXTURE GROUPS | | FIXTURE | ENTER TOTAL NUMBER OF FIXTURES | PROBABILITY OF USE (%) | ENTER FIXTURE FLOW RATE (GPM) | MAXIMUM RECOMMENDED FIXTURE FLOW RATE (GPM) |
|---|---|---|---|---|---|---|
| Bathroom Fixtures | 1 | Bathtub (no Shower) | 0 | 0.55 | 5.5 | 5.5 |
| | 2 | Bidet | 0 | 0.60 | 2.0 | 2.0 |
| | 3 | Combination Bath/Shower | 31 | 2.10 | 5.5 | 5.5 |
| | 4 | Faucet, Lavatory | 0 | 1.37 | 1.5 | 1.5 |
| | 5 | Shower, per head (no Bathtub) | 31 | 1.44 | 2.0 | 2.0 |
| | 6 | Water Closet, 1.28 GPF Gravity Tank | 31 | 0.60 | 3.0 | 3.0 |
| Kitchen Fixtures | 7 | Dishwasher | 23 | 0.37 | 1.3 | 1.3 |
| | 8 | Faucet, Kitchen Sink | 23 | 1.37 | 2.2 | 2.2 |
| Laundry Room Fixtures | 9 | Clothes Washer | 2 | 2.04 | 3.5 | 3.5 |
| | 10 | Faucet, Laundry | 1 | 1.37 | 2.0 | 2.0 |
| Bar/Prep Fixtures | 11 | Faucet, Bar Sink | 0 | 1.37 | 1.5 | 1.5 |
| Other Fixtures | 12 | Fixture 1 | 0 | 0.00 | 0.0 | 6.0 |
| | 13 | Fixture 2 | 0 | 0.00 | 0.0 | 6.0 |
| | 14 | Fixture 3 | 0 | 0.00 | 0.0 | 6.0 |

COMPUTED RESULTS FOR PEAK PERIOD CONDITIONS

Total No. of Fixtures in Calculation: n = 142

99th Percentile Demand Flow: Q = 19.8 GPM

Hunter Number: H(n,p) = 1.74

Stagnation Probability: Pr[Zero Demand] = 17%

DOWNLOAD RESULT | RESET WDC | ↓ Select Units for Water Demand ↓ GPM LPM LPS | RUN WDC

*Source:* Water Demand Calculator v2.1

**Figure 4-12(a)** WDC input and output for Denver site 114. Demand falls in the yellow region of Figure 4-10 and, hence, WDC uses the modified Wistort method to estimate peak demand. Results appear in Table 4-3.

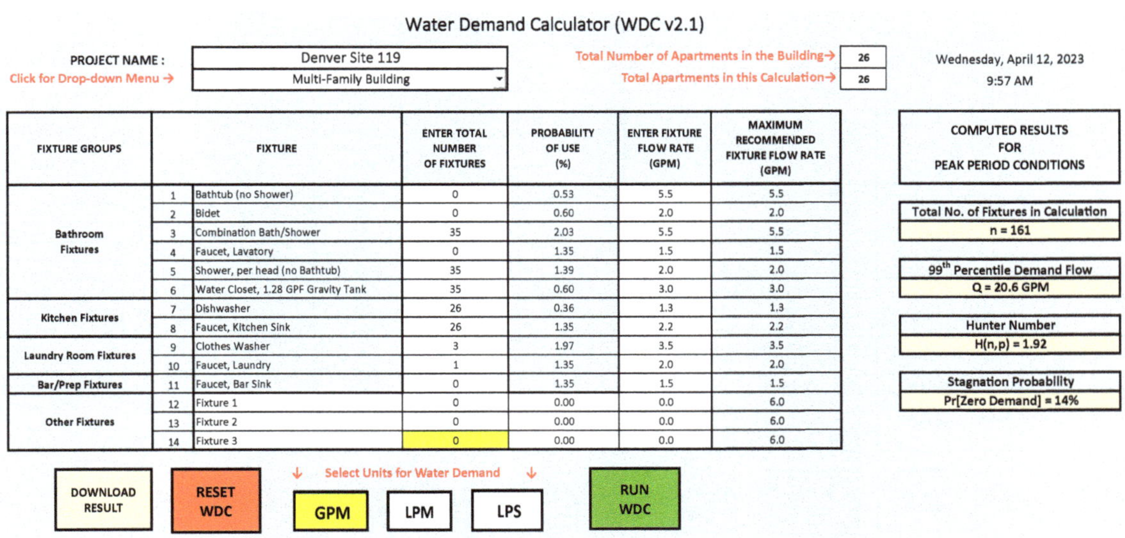

Water Demand Calculator (WDC v2.1)

PROJECT NAME : Denver Site 119

Click for Drop-down Menu → Multi-Family Building

Total Number of Apartments in the Building→ 26

Total Apartments in this Calculation→ 26

Wednesday, April 12, 2023

9:57 AM

| FIXTURE GROUPS | | FIXTURE | ENTER TOTAL NUMBER OF FIXTURES | PROBABILITY OF USE (%) | ENTER FIXTURE FLOW RATE (GPM) | MAXIMUM RECOMMENDED FIXTURE FLOW RATE (GPM) |
|---|---|---|---|---|---|---|
| Bathroom Fixtures | 1 | Bathtub (no Shower) | 0 | 0.53 | 5.5 | 5.5 |
| | 2 | Bidet | 0 | 0.60 | 2.0 | 2.0 |
| | 3 | Combination Bath/Shower | 35 | 2.03 | 5.5 | 5.5 |
| | 4 | Faucet, Lavatory | 0 | 1.35 | 1.5 | 1.5 |
| | 5 | Shower, per head (no Bathtub) | 35 | 1.39 | 2.0 | 2.0 |
| | 6 | Water Closet, 1.28 GPF Gravity Tank | 35 | 0.60 | 3.0 | 3.0 |
| Kitchen Fixtures | 7 | Dishwasher | 26 | 0.36 | 1.3 | 1.3 |
| | 8 | Faucet, Kitchen Sink | 26 | 1.35 | 2.2 | 2.2 |
| Laundry Room Fixtures | 9 | Clothes Washer | 3 | 1.97 | 3.5 | 3.5 |
| | 10 | Faucet, Laundry | 1 | 1.35 | 2.0 | 2.0 |
| Bar/Prep Fixtures | 11 | Faucet, Bar Sink | 0 | 1.35 | 1.5 | 1.5 |
| Other Fixtures | 12 | Fixture 1 | 0 | 0.00 | 0.0 | 6.0 |
| | 13 | Fixture 2 | 0 | 0.00 | 0.0 | 6.0 |
| | 14 | Fixture 3 | 0 | 0.00 | 0.0 | 6.0 |

COMPUTED RESULTS FOR PEAK PERIOD CONDITIONS

Total No. of Fixtures in Calculation: n = 161

99th Percentile Demand Flow: Q = 20.6 GPM

Hunter Number: H(n,p) = 1.92

Stagnation Probability: Pr[Zero Demand] = 14%

DOWNLOAD RESULT | RESET WDC | ↓ Select Units for Water Demand ↓ GPM LPM LPS | RUN WDC

*Source:* Water Demand Calculator v2.1

**Figure 4-12(b)** WDC input and output for Denver site 119. Demand falls in the yellow region of Figure 4-10 and, hence, WDC uses the modified Wistort method to estimate peak demand. Results appear in Table 4-3.

Water Demand Calculator (WDC v2.1)

PROJECT NAME : Sydney Building #1

Click for Drop-down Menu → Multi-Family Building

Total Number of Apartments in the Building→ 143

Total Apartments in this Calculation→ 143

Wednesday, April 12, 2023
11:48 AM

| FIXTURE GROUPS | | FIXTURE | ENTER TOTAL NUMBER OF FIXTURES | PROBABILITY OF USE (%) | ENTER FIXTURE FLOW RATE (GPM) | MAXIMUM RECOMMENDED FIXTURE FLOW RATE (GPM) |
|---|---|---|---|---|---|---|
| Bathroom Fixtures | 1 | Bathtub (no Shower) | 0 | 0.35 | 5.5 | 5.5 |
| | 2 | Bidet | 0 | 0.53 | 2.0 | 2.0 |
| | 3 | Combination Bath/Shower | 0 | 1.26 | 5.5 | 5.5 |
| | 4 | Faucet, Lavatory | 239 | 1.05 | 1.5 | 1.5 |
| | 5 | Shower, per head (no Bathtub) | 239 | 0.83 | 2.0 | 2.0 |
| | 6 | Water Closet, 1.28 GPF Gravity Tank | 239 | 0.53 | 3.0 | 3.0 |
| Kitchen Fixtures | 7 | Dishwasher | 143 | 0.30 | 1.3 | 1.3 |
| | 8 | Faucet, Kitchen Sink | 145 | 1.05 | 2.2 | 2.2 |
| Laundry Room Fixtures | 9 | Clothes Washer | 143 | 1.18 | 3.5 | 3.5 |
| | 10 | Faucet, Laundry | 143 | 1.05 | 2.0 | 2.0 |
| Bar/Prep Fixtures | 11 | Faucet, Bar Sink | 0 | 1.05 | 1.5 | 1.5 |
| Other Fixtures | 12 | Fixture 1 | 0 | 0.00 | 0.0 | 6.0 |
| | 13 | Fixture 2 | 0 | 0.00 | 0.0 | 6.0 |
| | 14 | Fixture 3 | 0 | 0.00 | 0.0 | 6.0 |

COMPUTED RESULTS FOR PEAK PERIOD CONDITIONS

Total No. of Fixtures in Calculation
n = 1,291

99th Percentile Demand Flow
Q = 42.2 GPM

Hunter Number
H(n,p) = 11

Stagnation Probability
Pr[Zero Demand] = 0%

DOWNLOAD RESULT | RESET WDC | Select Units for Water Demand: GPM | LPM | LPS | RUN WDC

*Source:* Water Demand Calculator v2.1

**Figure 4-13(a)** WDC output for a multifamily building in Sydney, Australia. This building falls in the red region of Figure 4-10 and, hence, WDC uses the Wistort method to estimate peak demand. Results appear in Table 4-4.

Water Demand Calculator (WDC v2.1)

PROJECT NAME : Canberra Building #1

Click for Drop-down Menu → Multi-Family Building

Total Number of Apartments in the Building→ 330

Total Apartments in this Calculation→ 330

Tuesday, May 16, 2023
11:31 AM

| FIXTURE GROUPS | | FIXTURE | ENTER TOTAL NUMBER OF FIXTURES | PROBABILITY OF USE (%) | ENTER FIXTURE FLOW RATE (GPM) | MAXIMUM RECOMMENDED FIXTURE FLOW RATE (GPM) |
|---|---|---|---|---|---|---|
| Bathroom Fixtures | 1 | Bathtub (no Shower) | 4 | 0.28 | 5.5 | 5.5 |
| | 2 | Bidet | 0 | 0.50 | 2.0 | 2.0 |
| | 3 | Combination Bath/Shower | 0 | 1.00 | 5.5 | 5.5 |
| | 4 | Faucet, Lavatory | 536 | 0.92 | 1.5 | 1.5 |
| | 5 | Shower, per head (no Bathtub) | 527 | 0.65 | 2.0 | 2.0 |
| | 6 | Water Closet, 1.28 GPF Gravity Tank | 525 | 0.50 | 3.0 | 3.0 |
| Kitchen Fixtures | 7 | Dishwasher | 330 | 0.28 | 1.3 | 1.3 |
| | 8 | Faucet, Kitchen Sink | 330 | 0.92 | 2.2 | 2.2 |
| Laundry Room Fixtures | 9 | Clothes Washer | 330 | 0.92 | 3.5 | 3.5 |
| | 10 | Faucet, Laundry | 330 | 0.92 | 2.0 | 2.0 |
| Bar/Prep Fixtures | 11 | Faucet, Bar Sink | 0 | 0.92 | 1.5 | 1.5 |
| Other Fixtures | 12 | Fixture 1 | 0 | 0.00 | 0.0 | 6.0 |
| | 13 | Fixture 2 | 0 | 0.00 | 0.0 | 6.0 |
| | 14 | Fixture 3 | 0 | 0.00 | 0.0 | 6.0 |

COMPUTED RESULTS FOR PEAK PERIOD CONDITIONS

Total No. of Fixtures in Calculation
n = 2,912

99th Percentile Demand Flow
Q = 71.5 GPM

Hunter Number
H(n,p) = 21

Stagnation Probability
Pr[Zero Demand] = 0%

DOWNLOAD RESULT | RESET WDC | Select Units for Water Demand: GPM | LPM | LPS | RUN WDC

*Source:* Water Demand Calculator v2.1

**Figure 4-13(b)** WDC output for a multifamily building in Canberra, Australia. This building falls in the red region of Figure 4-10 and, hence, WDC uses the Wistort method to estimate peak demand. Results appear in Table 4-4.

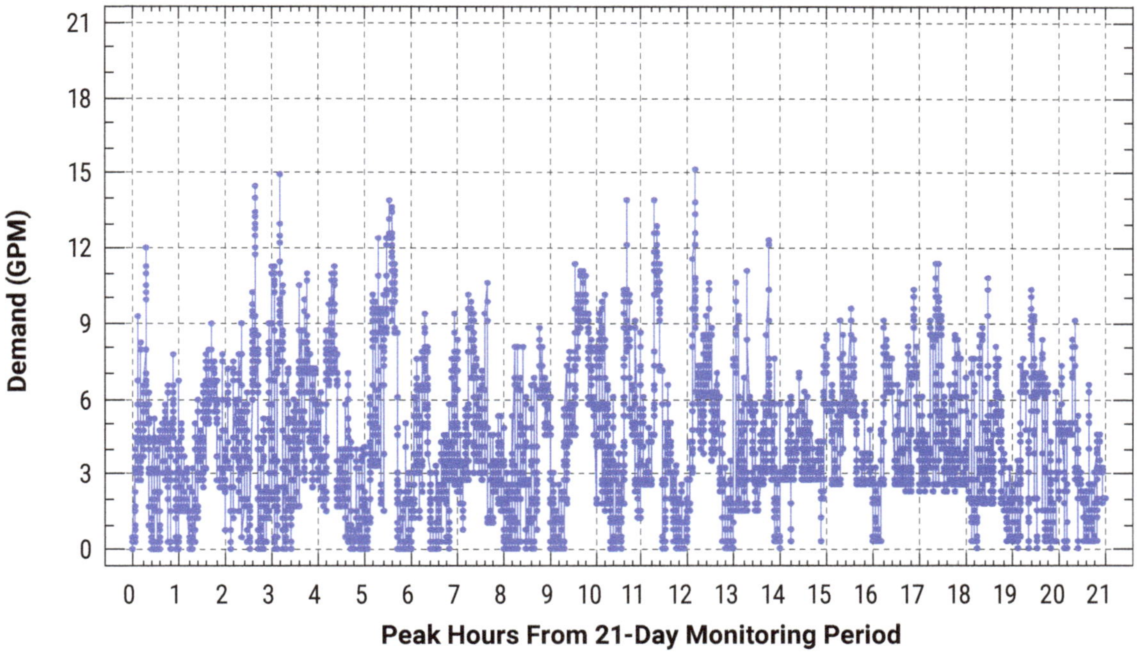


Figure 4-14(a) Water demands measured during the peak hour at Denver Building 119 collected every 10 seconds over a 21-day period

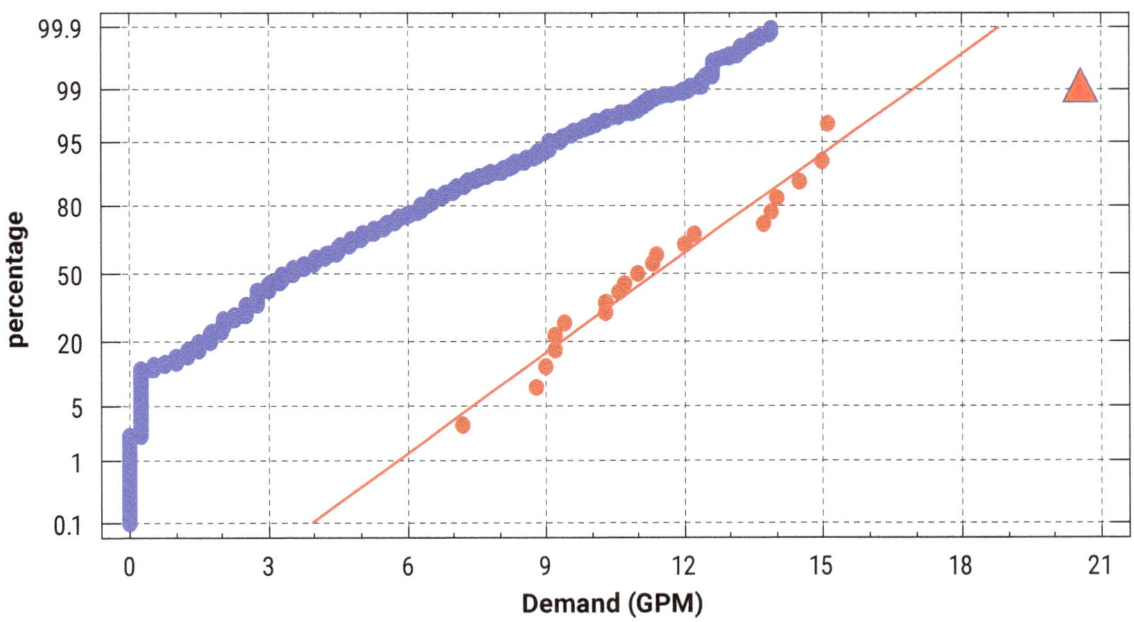


Figure 4-14(b) Normal probability plots of two data sets from Denver Building 119. The blue data set (7,560 points) is from Figure 4-14(a). The red data set (21 points) is the maximum values measured during each of the 21 peak hourly periods. The triangle is the 99th percentile demand predicted by the WDC.

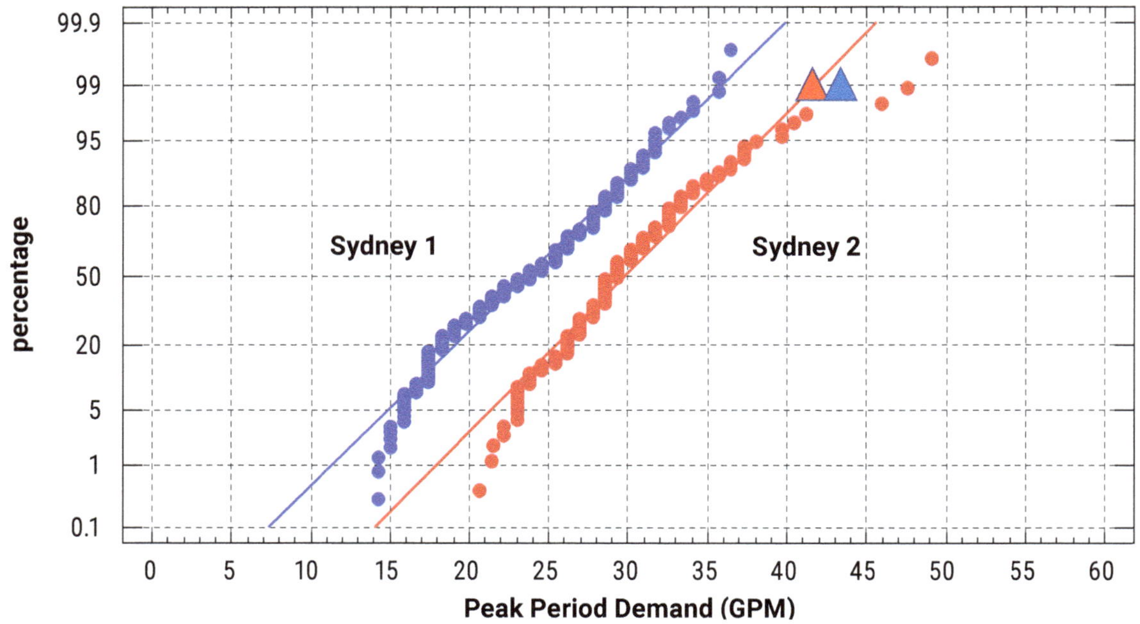


Figure 4-15(a) Normal probability plots of daily peak demands at two multifamily buildings in Sydney, Australia. The two triangles represent the 99th percentile demand predicted by the WDC (see values in Table 4-4 and Figure 4-13[a]).

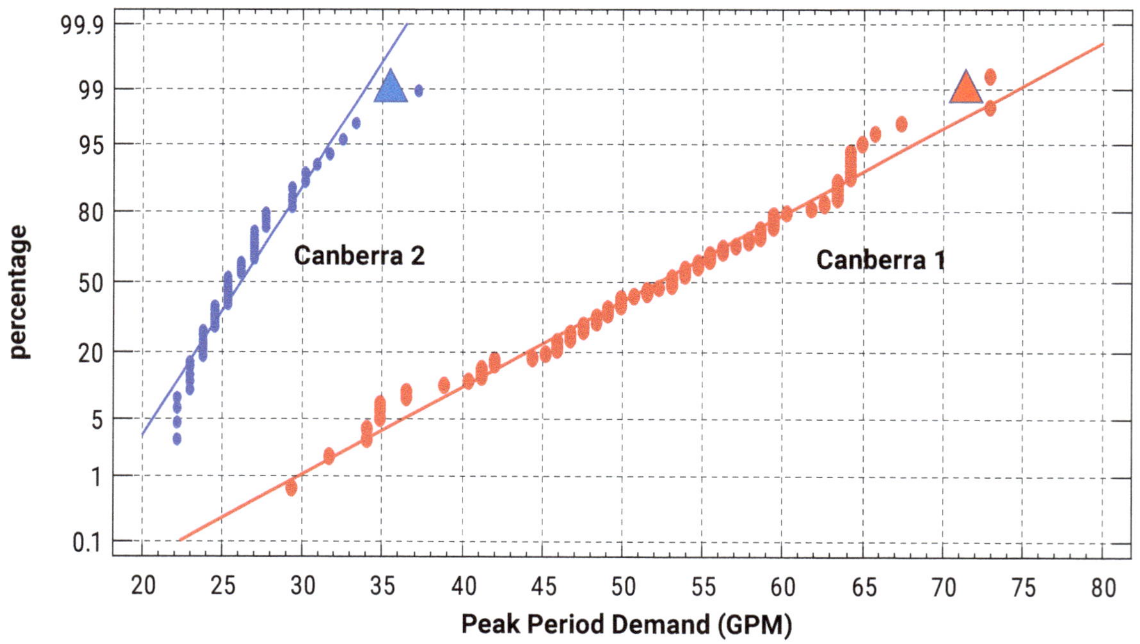


Figure 4-15(b) Normal probability plots of daily peak demands at two multifamily buildings in Canberra, Australia. The two triangles represent the 99th percentile demand predicted by the WDC (see values in Table 4-4 and Figure 4-13[b]).

# ESTIMATING PEAK DEMAND FOR OTHER CASES

## Introduction to NFPA 13D and Domestic-fed Multipurpose System Meter Sizing

The National Fire Protection Association (NFPA) 13D Standard for the Installation of Sprinkler Systems in One- and Two-Family Dwellings and Manufactured Homes was adopted in 1975 to reduce annual loss of life due to residential fires. Single-family homes, two-family dwellings, and manufactured homes or townhomes can use these fire-protection systems to reduce the chance of flashovers and improve chances of evacuation during a fire event. Additionally, please refer to the current version of the *International Fire Code*, specifically Chapter 9, for additional information on requirements that apply to the design and installation of fire protection and life safety systems. NFPA 13D fire systems are not approved for commercial building or multifamily residential buildings, including low-rise (four stories or less) residential multifamily buildings. Per the NFPA 13D, multipurpose piping configurations and one single-tapped, stand-alone piping configuration (see lower-right graphic in Figure 6-3 in Chapter 6 of M22) serving both domestic water fixtures and fire sprinkler systems through a single tap and service line are permissible. There are several acceptable piping alternatives presented in the NFPA 13D Standard showing these multipurpose sprinkler systems and other stand-alone sprinkler systems. Please refer to the schematic diagrams shown in Chapter 6 of this manual and the NFPA 13D Standard for more specific information on sprinkler design and required system components. Additionally, see the publication *Residential Fire Sprinkler Systems: Guidance for Water Utilities* (AWWA 2018) for a more detailed overview and history of NFPA 13D and its use in residential sprinkler design.

## Basic NFPA 13D Information and Water Meter Sizing

When sizing water meters for domestic-fed multipurpose or single-tapped stand-alone systems, it is important to not only consider the maximum design flow of the combined domestic and fire system but also the accuracy of the meter to measure low flow demands. It is typical for utilities to require larger water meters because of the possibility of higher design flows when combining domestic demand, irrigation demand, and fire flow demand. As implied by the title of the NFPA 13D Standard, the water meter may be sized for either a single-family residence or an attached multifamily structure such as townhomes with independent piping systems. The pressure and flow of a residential sprinkler system is determined by the specifications of the sprinkler head and NFPA requirements. The different discharge coefficients drive the amount of water that is permitted through the sprinkler, and this can range from 8 gpm to 20 gpm per sprinkler head depending on the system design (Johnson Controls 2021). Typically, the minimum design demand for a residential sprinkler is around 13 gpm with a minimum design pressure based on the NFPA 13D Standard of 7 psi. Based on NFPA 13D requirements that require hydraulic calculations to include at least two sprinkler heads, the minimum value for a fire sprinkler system for a single residential dwelling is typically 26 gpm. This demand may increase depending on building type (single-family residential versus attached multifamily structures), outdoor water demands (irrigation systems), sprinkler head requirements, and whether the authority having jurisdiction has standards exceeding the NFPA 13D requirements. Because of varying local requirements and the increasing sizes of homes with multiple dwelling units, there may be certain instances where a NFPA 13R system might be a more economical or otherwise a more favored option by the designer/engineer.

When calculating the peak demand of the sprinkler system serving more than one dwelling unit, such as duplexes and townhomes, the NFPA 13D Standard does not require the summation of the total fire flow demand of each sprinkled dwelling units. Instead, NFPA 13D calls for the base sprinkler demand of a single dwelling unit with the

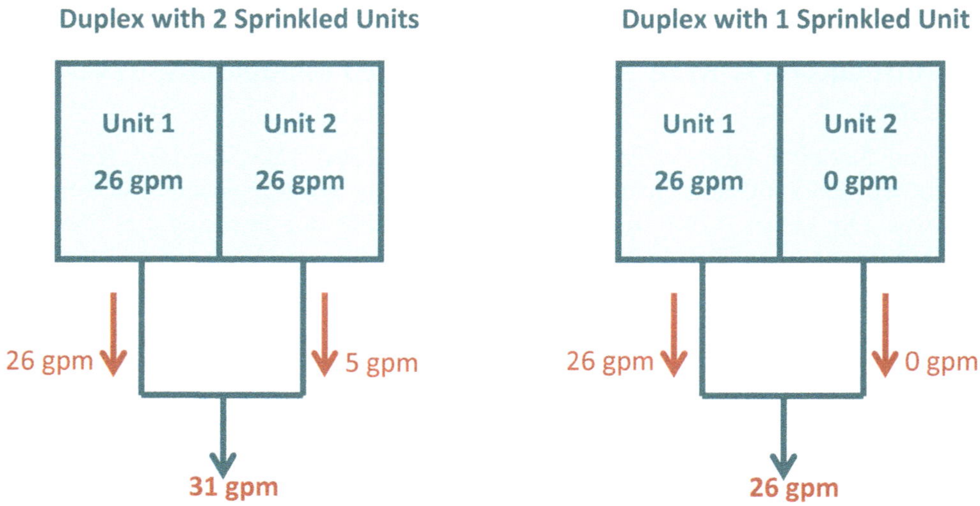


Figure 4-16 Common supply fire demand calculation per NFPA 13D, with more than one dwelling

addition of 5 gpm for each additional dwelling unit. As a best practice, the base sprinkler demand should represent the maximum required NFPA 13D flow for the largest single unit. Figure 4-16 illustrates the concept of calculating the fire demand for an attached multifamily structure with independent internal plumbing. Depending on the building code driving the construction of the dwelling (either *The International Building Code* or *The International Residential Code)*, sprinkler protection may only be necessary if fire access is not satisfied. In these cases, where sprinklers are required due to fire access, it is likely not all the residential dwelling units may require a sprinkler system.

## Water Meter Sizing Guidelines and Examples

Based on varying statewide adoptions of the NFPA 13D Standard and jurisdictional requirements, there are several variations on how to determine water meter size for a domestic-fed multipurpose or single-tapped stand-alone system. This section presents some examples for sizing water meters in three jurisdictions: East Bay Municipal Utility District (EBMUD), Denver Water (DW), and the City of Scottsdale, Ariz.

**East Bay Municipal Utility District.** EBMUD permits multipurpose systems for one- and two-family dwellings under California State Building Code. EBMUD sets a standard service connection (lateral) size of 1.5 in. with a minimum meter size of 1 in. The meter size is based on the larger of two calculated demands: the total domestic peak demand based on a fixture unit count (including both indoor and outdoor water use) or the calculated sprinkler demand plus a 15-gpm domestic allowance (East Bay Municipal Utility District, 2019).

**Denver Water.** DW uses a different method for sizing domestic-fed multipurpose system meters. The water meter size is calculated by comparing the indoor domestic peak demand to the automatic fire sprinkler demand. The meter is sized using indoor domestic peak demand, and the tap and service line are sized based on the fire sprinkler demand. If there is a greater than one size difference between the sizing of the meter using the domestic peak demand and the sizing of the tap and service line using the fire sprinkler demand, the dwelling will be required to use the version of stand-alone sprinkler system design where the domestic and fire services are tapped independently of each other, as shown in Figure 6-2 in Chapter 6 (Denver Water 2019). At the time of publication, DW is considering adopting the Water Demand Calculator for use in multifamily residential meter sizing.

**City of Scottsdale, Ariz.** Scottsdale has a specific code for sizing water meters for a domestic-fed multipurpose system for both new construction and existing residential structures undergoing additions and alterations. Per the City of Scottsdale's *Interpretations and Applications of NFPA 13D* (City of Scottsdale 2019b) and building code, domestic service lines are set at a minimum of 1 in. for newly constructed residential homes. Water meters are sized to meet calculated demands based on the adopted city building code. Existing residential structures that are being upsized or undergoing renovations may use their existing ⅝-in. or ¾-in. meter if hydraulic calculations show that ample water supply can be delivered to the fire sprinkler system. Larger single-family homes over 12,000 ft$^2$ are required to have a minimum 1.5-in. domestic meter and a minimum 2-in. domestic service line. Single-family homes with additional challenges including access and size may not qualify for a 13D system (City of Scottsdale 2019b).

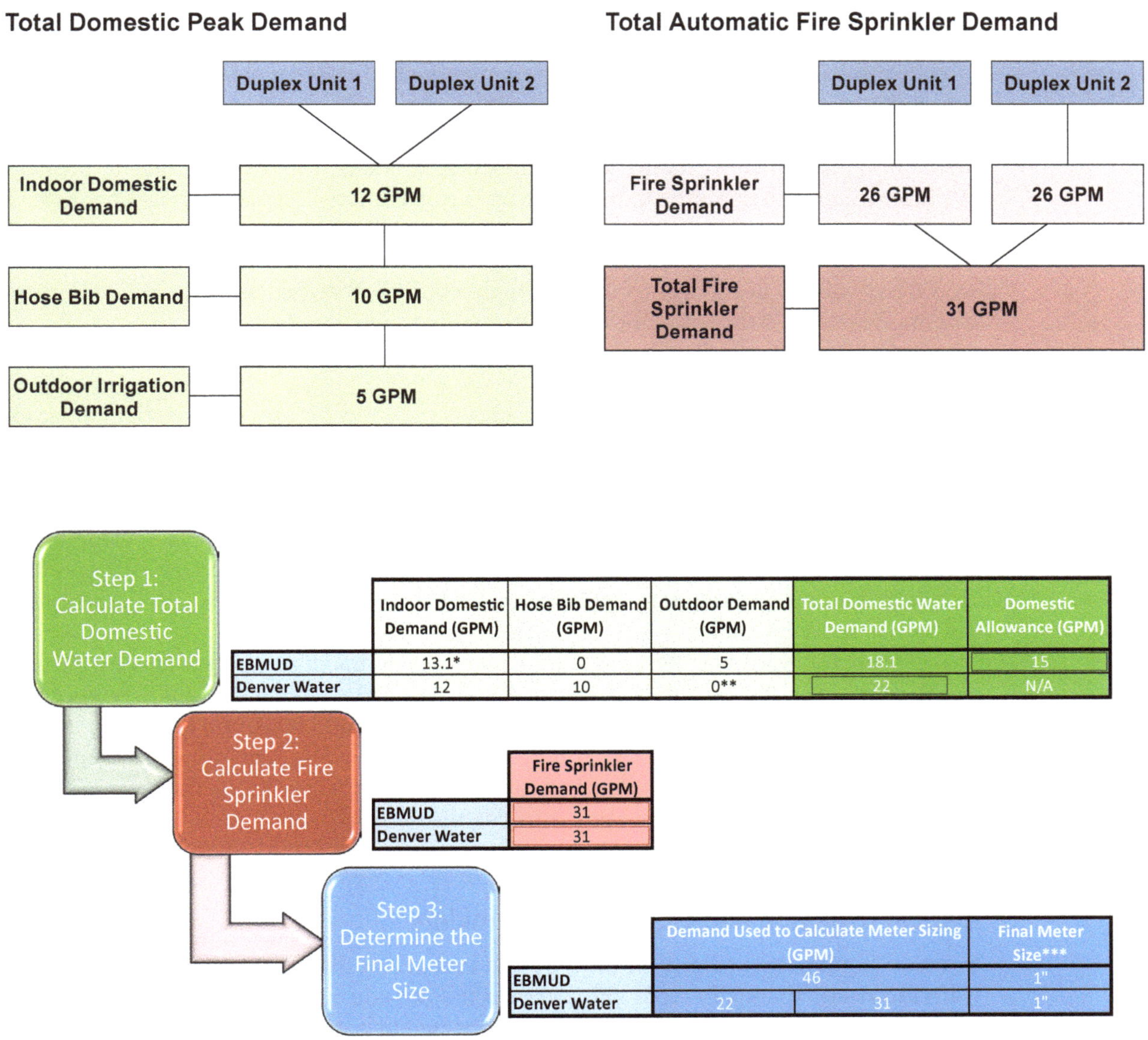


| | Indoor Domestic Demand (GPM) | Hose Bib Demand (GPM) | Outdoor Demand (GPM) | Total Domestic Water Demand (GPM) | Domestic Allowance (GPM) |
|---|---|---|---|---|---|
| EBMUD | 13.1* | 0 | 5 | 18.1 | 15 |
| Denver Water | 12 | 10 | 0** | 22 | N/A |

| | Fire Sprinkler Demand (GPM) |
|---|---|
| EBMUD | 31 |
| Denver Water | 31 |

| | Demand Used to Calculate Meter Sizing (GPM) | | Final Meter Size*** |
|---|---|---|---|
| EBMUD | 46 | | 1" |
| Denver Water | 22 | 31 | 1" |

*Hose bib demand is calculated based on fixture units when calculating indoor domestic demand per EBMUD meter sizing documentation.
**Irrigation is assumed to occur at off peak hours between 1AM and 4AM and is therefore not included (per DW standards)
***Meter size based on current engineering and sizing standards located on the utility website.

**Figure 4-17 Domestic-fed multipurpose sizing example**

The water meter sizing example calculation shown in Figure 4-17 is for a duplex residence using a domestic-fed multipurpose system with a single water meter. The example shows the summation of the domestic demand and fire sprinkler demand for the duplex and how the water meter size is calculated at DW and EBMUD. For DW, the higher fire sprinkler flow of 31 gpm determined the size of the meter.

In the example, although meter sizing was determined differently, both utilities require the same meter size. Please refer to the AWWA meter standards information in Table 6-1 of this M22 manual for information on normal operating ranges of water meters. There are several variations on how the authority having jurisdiction manages meter flow rate standards, velocity, and head loss when sizing water meters. It is important to ensure when designing water systems and sizing meters that three points are followed: (1) there is a full understanding of what is allowed by the local water utility; (2) recent updates to the NFPA 13D Standard are obeyed; and (3) local building and fire codes are followed.

## LANDSCAPE IRRIGATION DEMAND

Turf irrigation is one of the largest water demands that utility customers may encounter, especially in arid climates. Large amounts of water in excess of domestic use are often required for turf irrigation, thereby necessitating particular care in calculating the irrigation demand. If irrigation demand is underestimated, there can be insufficient pressure available to operate domestic fixtures during turf irrigation. A separate irrigation meter is often desirable because wastewater charges are typically based on building water use. If that use includes irrigation (which does not contribute to wastewater flows), a customer would pay wastewater charges on a much higher volume than necessary. Using a dedicated irrigation meter will track specific irrigation demands and ensure that unnecessary charges for wastewater are avoided. Some utilities compute wastewater charges during a winter month when there is no irrigation use.

In California, a separate irrigation meter is required for any commercial landscape larger than 5,000 ft$^2$. Single-family residential services are exempted from this requirement. There is no harm in installing a dedicated irrigation meter with an approved backflow prevention device at sites where irrigation is practiced, regardless of the landscape size. The rule of thumb established through previous editions of M22 is to recommend a separate meter serving irrigation for any landscape (commercial or residential) larger than 15,000 ft$^2$. Requiring a separate irrigation meter for smaller landscapes, as is practiced in California, is acceptable.

Automatic sprinkler systems should be programmed to operate during periods of off-peak domestic demands (from late night to early morning). This is even more important if a single meter is used to serve both domestic and irrigation demands. It is not recommended to simply add the full irrigation demand to the peak domestic demand to calculate the maximum design demand for the customer. Therefore, where the irrigation system is governed by an automatic controller, the design demand should be the larger of the domestic demand or the irrigation demand.

### Manual Irrigation and End-of-hose Sprinklers

In many homes, particularly in older residential areas, the resident supplies the lawn-watering needs manually from sill cocks on the outside wall of the building and from yard-hose bibbs. The sill cocks usually are connected to a ½ -in. or ¾ -in. supply pipe, and the use of these outlets reduces water pressures within the house. Hoses that connect the sprinkler to the hose bibb vary from ⅜ in. to ¾ in. in diameter and govern the rate of flow due to friction loss. Sprinklers can be obtained in a wide range of water-flow capacities and types, which makes detailed estimation of peak demand difficult.

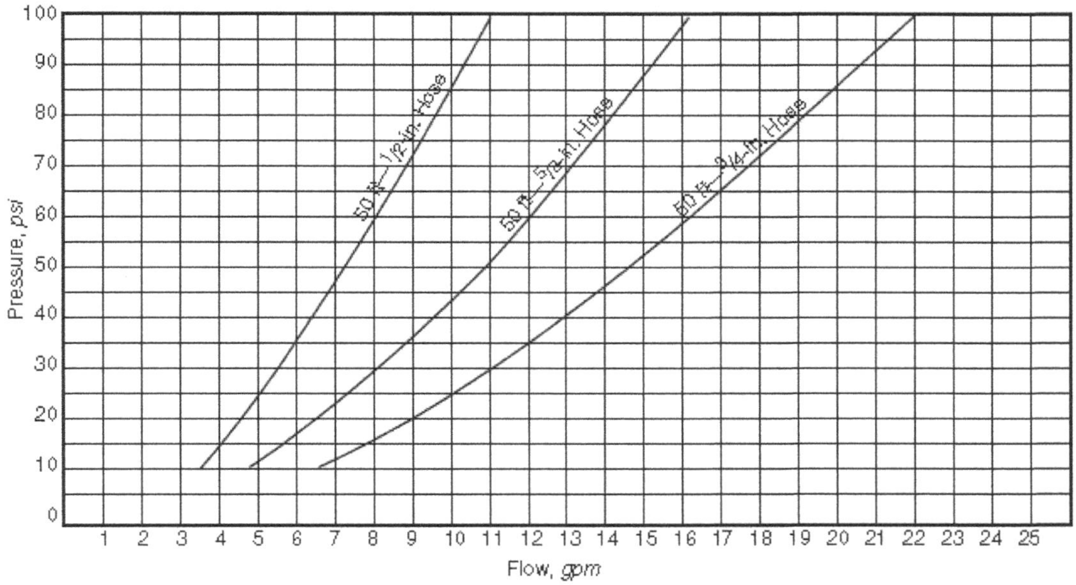


**Figure 4-18 Variations in flow from garden hoses due to pressure changes**

For manual irrigation systems, the demand from at least one garden hose should be included as part of the peak demand. Results of flow tests on a ring sprinkler provide guidance on portable-sprinkler water conditions at various pressures. For example, Figure 4-18 shows the flow variation through 50 ft of ⅝-in. diameter hose and a brass ring sprinkler fed by a 1½-in. house service line and a ¾-in. service line to a ¾-in. sill cock.

## Basic Line Sizing for Automatic Irrigation Systems

The Irrigation Association (IA) recommends that the flow velocity in an automatic irrigation system not exceed 7.5 ft/s (IA 2012, 2014). This requirement effectively limits the area of landscape that can be watered through a given service line. There are several approaches for sizing dedicated irrigation service lines. The method presented here was developed in conjunction with IA and offers a rule-of-thumb approach to size an irrigation meter based on landscape area and service line diameter. While not intended as a replacement for customized meter-sizing methods, the IA approach is useful to confirm right sizing of irrigation meters and service lines. The peak daily demand is determined and then adjusted based on the chosen irrigation interval (e.g., every 2 days, 3 days) and watering window during which irrigation must be completed. As the length of the irrigation interval increases, the size of the water tap also increases. The flow rate required to provide sufficient water to meet the peak day irrigation demand is calculated using Eq. 4-6.

$$Q = (0.623)\left[\frac{ET_0 \times K_L \times I_N \times A}{W \times IE}\right] \qquad \text{(Eq. 4-6)}$$

Where:

$Q$ = Flow of water (gpm)
$ET_0$ = Grass reference evapotranspiration (in./day; use peak demand day)
$K_L$ = Landscape coefficient to modify $ET_0$ (for planning purposes, use 0.70)
$I_N$ = Irrigation interval (days; for planning purposes, use 2 days)
$A$ = Area of landscape to be irrigated ($ft^2$)
W = Watering window to accomplish irrigation for peak demand (minutes)
$IE$ = Irrigation efficiency (for planning purposes, use 0.75).

In Eq. 4-6, the constant 0.623 is the factor to convert (in.·ft$^2$) to (gal). The result gives the flow rate in gpm to irrigate the landscape area during peak demand based on the specified watering window and interval for watering days. Irrigation efficiency is determined by comparing water beneficially used to water applied. This accounts for sprinkler distribution uniformity and irrigation management skill. It also accounts for water that is delivered but lands off target or is applied excessively and becomes runoff or deep percolation. It is suggested that 0.70–0.80 be used as the typical irrigation efficiency for permanently installed sprinklers. While the use of drip irrigation will have a higher potential efficiency, for planning purposes a target irrigation efficiency of 0.75 is acceptable. For reference, the California Model Efficient Landscape Water Ordinance uses $IE = 0.71$ for irrigation efficiency.

Rearranging Eq. 4-6 gives the landscape area, A, in square feet as shown in Eq. 4-7, where all terms have been previously defined. This is the area that can be served by a meter and service line with a designated flow capacity.

$$A = \left(\frac{1}{0.623}\right)\frac{Q \times W \times IE}{ET_0 \times K_L \times I_N} \qquad \text{(Eq. 4-7)}$$

Assuming a 100 percent turf landscape in a hot and dry region, to be watered by overhead irrigation including spray or rotor-type sprinklers, it is possible to develop reasonable estimates of the maximum landscape size that can be served through service lines of different sizes. Table 4-5 shows the results of this analysis. Typically, water meters connected to these service lines will be the same size or one size smaller than the service line. Note that the results in Table 4-5 provide general guidance and are not intended to replace a customized meter and line sizing calculation that is site-specific. Rather, Table 4-5 provides a rule-of-thumb reality check on irrigation meter sizing that can be used to help water providers ensure proper flow and velocity within an irrigation system.

Landscaped areas larger than those listed in Table 4-5 can be served if irrigation system flows are lowered. Flows can easily be lowered through the use of drip and micro-spray as well as low-precipitation rate sprinkler heads. Key assumptions in this table include the $ET_0$ that is set at 0.5 in. per day and $I_N$, the irrigation interval that is set to every two days. These conditions reflect a very hot and dry climate, where 0.5-in. of supplemental

**Table 4-5** Service line size and maximum irrigated turf area for watering window of $W = 840$ min (14 h) and irrigation interval of $I_N = 2$ days.

| Service Line Size (in.) | Q = Maximum Recommended Continuous Flow Rate (gpm)* | $ET_0$ (high) = Maximum Irrigation Requirement Based on Climate (in./day) | Maximum Irrigated Turf Area (ft$^2$)† | Maximum Irrigated Turf Area (Acres)† |
|---|---|---|---|---|
| ⅝ | 7 | 0.5 | 10,110 | 0.2 |
| ¾ | 10 | 0.5 | 14,450 | 0.3 |
| 1 | 17 | 0.5 | 24,560 | 0.6 |
| 1.25 | 27 | 0.5 | 39,000 | 0.9 |
| 1.5 | 37 | 0.5 | 53,450 | 1.2 |
| 2 | 65 | 0.5 | 93,900 | 2.2 |
| 2.5 | 100 | 0.5 | 144,460 | 3.3 |
| 3 | 155 | 0.5 | 223,920 | 5.1 |

* Based on recommendation from the Irrigation Association to ensure flow velocity does not exceed 7.5 ft/s.

† Assumes 100 percent turf and overhead irrigation to include spray or rotor-type sprinklers. Larger landscape areas can be served if irrigation system flows are lowered with drip and micro-spray.

irrigation water is required every two days (1.5 in. of irrigation per week). More climates in North America will approach this previously "worst case" sizing scenario as temperatures increase. Adjusting the factors assumed here for local conditions allows this method to be tailored to meet the specific demands of an irrigation system in a particular climate zone.

## REFERENCES AND WEBSITES


American Water Works Association (AWWA). 2014 (3rd ed.). M22, *Sizing Water Service Lines and Meters*. Denver, Colo: AWWA.

AWWA. 2004 (2nd ed.). M22, *Sizing Water Service Lines and Meters*. Denver, Colo: AWWA.

AWWA. 2018. *Residential Fire Sprinkler Systems: Guidance for Water Utilities*. Denver, Colo.: AWWA.

Benazzi, R.V.; Guss, L.; Orend, R.J.; Wyly, R.S.; and Konen, T.P. 1976. "Water Requirements and Procedures for Estimating the Demand for Water in Buildings." Proc. CIBW62 Symposium, Washington, DC.

Buchberger, S.G.; Blokker, E.M.J.; and Cole., D.P. 2012. "Estimating Peak Water Demands in Hydraulic Systems I – Current Practice." Proc. WDSA2012 Symposium, Adelaide, South Australia, Sept. 24-27, 2012.

Buchberger, S.; Omaghomi, T.; Wolfe, T.; Hewitt, J.; and Cole, D. 2017. "Peak Water Demand Study – Probability Estimates for Efficient Fixtures in Single & Multi-Family Residential Buildings" (Exec. Summary). Ontario, Calif.: IAPMO.

California Model Efficient Landscape Water Ordinance. https://water.ca.gov/Programs/Water-Use-And-Efficiency/Urban-Water-Use-Efficiency/Model-Water-Efficient-Landscape-Ordinance.

City of Scottsdale. 2019a. *City of Scottsdale Code of Ordinances*. Retrieved from Codes & Ordinances: www.scottsdaleaz.gov/codes.

City of Scottsdale. 2019b. "Interpretations and Applications of NFPA 13D".

Cole, D. 2012. "Determining Fixture Units for High-Efficiency Fixtures." Ontario, Calif.: IAPMO.

DIN 1988-300. 2012. Codes of Practice for Drinking Water Installations: Pipe Sizing; DVGW Code of Practice. Berlin: Deutsches Institut für Normung e.V.

Douglas, C. 2019. "Instantaneous Water Demand Estimates for Buildings with Efficient Fixtures." MS Thesis. Cincinnati, Ohio: University of Cincinnati.

East Bay Municipal Utility District. (2019, September 13). *Applying for dual service*. Retrieved from East Bay Municipal Utility District: https://www.ebmud.com/customers/new-meter-installation/standard-water-service/applying-dual-service-single-family-and-duplex/

East Bay Municipal Utility District. (2019, Sept. 13). *Meter Sizing for Residential Fire Sprinklers*. Retrieved from East Bay Municipal Utility District.

Hobbs, I.; Anda, M.; and Bahri, P.A. 2019. "Estimating Peak Water Demand: Literature Review of Current Standing and Research Challenges." *Results in Engineering*. Vol. 4, Article 100055.

Hunter, R. 1940. BMS-65, "Methods of Estimating Loads in Plumbing Systems." Washington, DC: National Bureau of Standards (NBS).

Hunter, R. 1941. BMS-79, "Water Distribution Systems for Buildings." Washington, DC: US Government Publishing Office.

Hunter, R. 1946. BMS-66, "Plumbing Manual: Report of Subcommittee on Plumbing, Central Housing Committee on Research, Design, and Construction." Washington, DC: NBS.

International Association of Plumbing and Mechanical Officials (IAPMO). 2018a. *National Standard Plumbing Code*. Ontario, Calif.: IAPMO.

IAPMO. 2018b. *Uniform Plumbing Code*. Ontario, Calif.: IAPMO.

Irrigation Association (IA). 2012. Standard for Automatic Irrigation Systems. Falls Church, Va.: IA.

IA. 2014. Landscape Irrigation Best Management Practices. Falls Church, Va.: IA.

Jack, L. and Whorlow, B. 2017. "LUNA Project: An Assessment of the Validity of the Loading Units Method for Sizing Domestic Hot and Cold Water Services." London: Chartered Institution of Building Services Engineers (CIBSE).

Johnson Controls. 2021. *Tyco*. Retrieved from 4: Residential Sprinklers: https://docs.johnsoncontrols.com/tycofire/v/u/Tyco/en-US/Series-LFII-Residential-Sprinkler-Design-Guide.

Mayer, P.; Davis, S.E.; Buchberger, S.G.; and Douglas, C. 2020. Assessing Water Demand Patterns to Improve Sizing of Water Meters and Service Lines. Project 4689. Denver, Colo.: The Water Research Foundation.

National Fire Protection Association, Codes and Standards. www.nfpa.org. NFPA13D, accessed Nov. 13, 2019.

Omaghomi, T. and Buchberger, S.G. 2014. "Estimating Water Demands in Buildings." Proc. 16th Annual International Symposium on WDSA2014. Bari, Italy, July 14-17, 2014.

Omaghomi, T. and Buchberger, S.G. 2018. "Variation in Peak Water Demand with Building Size: Parameters and Methods." Proc. 1st International Joint CCWI-WDSA Conf., July 2018, Kingston, Ont.

Omaghomi, T. 2019. "Estimating Peak Water Demand in Buildings with Efficient Fixtures: Methods, Merits, and Implications." Ph.D Dissertation. Cincinnati, Ohio: University of Cincinnati.

Omaghomi, T.; Buchberger, S.G.; Cole, D.; Hewitt, J.; and Wolfe, T. 2020. "Probability of Water Fixture Use During Peak Hour in Residential Buildings." ASCE *Journal of Water Resources Planning and Management*. 146(5).

Steele, A. 1982. *Engineered Plumbing Design* (2nd ed.). Elmhurst, Ill.: Construction Industry Press.

USNCCIB (US National Committee for the International Council for Building Research, Studies, and Documentation Counterpart Commission to CIB W-62). 1974. "Water Distribution and Supply within Buildings: National Research Needs." Washington, DC: National Academy of Sciences.

*Water Plan Submittal—Step-By-Step Process for Plan Review*. https://www.denverwater.org/contractors/construction-information/plan-reviews/water-plan-submittal

Wistort, R.A. 1995. "A New Look at Determining Water Demands in Building: ASPE Direct Analytic Method." Proc. American Society of Plumbing Engineers Convention, Kansas City, Mo. 17-34.

**Websites** The following websites were visited between Dec. 2, 2019 and Jan. 31, 2020 to obtain fixture flow rate information and other specifications listed in Table 4-1.

- https://www.epa.gov/sites/production/files/2017-01/documents/ws-commercial-webinar-slides-kitchens.pdf
- https://www.epa.gov/sites/production/files/2017-01/documents/ws-products-factsheet-toilets_0.pdf
- https://19january2017snapshot.epa.gov/www3/watersense/commercial/docs/watersense_at_work/files/assets/basic-html/toc.html
- https://www.americanstandard-us.com
- https://www.sloan.com
- https://www.deltafaucet.com
- https://www.us.kohler.com
- https://www.moen.com/
- https://www.totousa.com/
- https://www.flushmate.com
- https://www.chicagofaucetshoppe.com
- https://www.webstaurantstore.com
- https://www.tsbrass.com/
- https://www.supplyworks.com
- https://www.bradleycorp.com/
- https://speakman.com/

- https://www.symmons.com/
- https://www.chicagofaucets.com/
- https://www.ferguson.com

## ACRONYMS AND ABBREVIATIONS

| | |
|---|---|
| ASCE | American Society of Civil Engineers |
| ASHRAE | American Society of Heating, Refrigerating and Air-Conditioning Engineers |
| ASPE | American Society of Plumbing Engineers |
| AWWA | American Water Works Association |
| CF | Cumulative frequency |
| ExEn | Exhaustive Enumeration |
| FU | Fixture unit |
| FV | Fixture value |
| IAPMO | International Association of Plumbing and Mechanical Officials |
| ICC | International Code Council |
| IPC | International Plumbing Code |
| IRC | International Residential Code |
| gpm | Gallons per minute |
| MWM | Modified Wistort method |
| M22 | AWWA M22, Sizing of Water Service Lines and Meters |
| NFPA | National Fire Protection Association |
| NSPC | National Standard Plumbing Code |
| RF | Relative frequency |
| SMOC | Safe Maximum Operating Capacity |
| UPC | Uniform Plumbing Code |
| WDC | Water Demand Calculator |
| WQA | Water Quality Association |
| WQRF | Water Quality Research Foundation |
| WRF | Water Research Foundation |
| WSFU | Water supply fixture unit |
| WM | Wistort method |

This page intentionally blank.

Chapter 5

# Service-Line Sizing

## INTRODUCTION

This chapter describes how to determine the size of water service lines and valves given the design flow rate through the service line and head available at the main. The service line is the pipe that connects the water main to the water meter. The determination of service line size is more conservative than the determination of meter size because the service line size must accommodate local water system operational limitations. In addition, there is a substantial cost associated with replacing a service line. The methodologies described cover a wide array of problems from sizing a single-family residence's service line to sizing large industrial and commercial services with fire hydrants and sprinklers. This chapter will not describe the sizing of customer plumbing, although some consideration of a customer's plumbing is necessary, especially when the system includes sprinklers and hydrants.

All customer situations have similar challenges. Only a limited amount of head is available in the main. Selecting larger-diameter service lines and appurtenances results in greater capacity services but also results in higher costs than the costs of smaller lines. The goal in design is to correctly size the combination of pipes, valves, and meter to meet the customer's needs based on the best available data. Travel time in service lines should be short so that water quality transformations should be minimal.

There are several ways of performing the hydraulic calculations for service line sizing, ranging from hand calculations to spreadsheets to sophisticated computer models. All these approaches require knowing the head at the main connection and then using some description of the relationship between flow rate and head loss to calculate the flow through the service line. Each piece of data required is described in an individual section in this chapter followed by a description of how to perform the calculations.

# DATA REQUIRED

## Design Flow Rate

The starting point for sizing calculations is the design flow rate. This rate can be estimated using the methods described in Chapter 4 unless local standards supersede the methods in the manual. Either of these design flow rates can be used in the subsequent calculations.

For the design of residential fire sprinkler systems (RFSS) in a stand-alone configuration (dedicated premise piping to sprinkler heads), both the local plumbing and/or fire code and the National Fire Protection Association's 13D standard (NFPA 2019c) should be consulted for the design flow rate. For a multipurpose system (single-premise piping that supplies domestic fixtures and sprinkler heads), the design flow rate for the service line is based on either the number of simultaneously activated sprinkler heads (a minimum of two heads at 13 gpm each totaling 26 gpm per the 2019 NFPA 13D standard) plus any additional flow requirement prescribed by local plumbing and/or fire or code, whichever is greater. Some utilities incorporate additional flow from outdoor irrigation systems, bringing the design flow to 40 gpm or higher. The NFPA 13D standard requires a minimum operating pressure of 7 psi at any sprinkler head. Additional design criteria for RFSS applications are given in Chapter 6.

## Head

Two heads (hydraulic grade line elevations) are used in service line sizing. The first is the head available in the main at the tap location. This head is composed of the pressure in the main combined with the elevation corresponding to that pressure. The second head is the required head at the downstream end of the service line. This head is composed of the minimum acceptable pressure and the elevation corresponding to that pressure. The difference between the available head at the main and the required head is the acceptable head loss in the service line, meter, and valving.

## Elevation of Main and Service Line

The elevation of the main is either the actual elevation of the main or the elevation of the pressure gauge attached to a nearby fire hydrant if the pressure used as the water utility pressure is measured at a hydrant. Similarly, the elevation of the customer service line is usually taken as the elevation immediately downstream of the meter, although for some situations the elevation of a particular hydrant or sprinkler on the customer's side of the meter is used. The distinction depends on whether the individual performing service line calculations is interested in only sizing the service lines or is sizing the complete customer plumbing system.

It is not as important to know the actual elevation of the main. It is important to know the difference in elevations between the two head points of interest. In areas that are flat (e.g., the elevation of the main is comparable to that of the meter), the difference in elevation may be negligible, and it may be possible to simply compare the difference in pressure between the main and points along the customer service line.

## Available Pressure

The available pressure, or head, used to start the calculation is typically taken as the working pressure under a reasonable worst-case condition (e.g., peak hour demands, fire demands, condition when nearby pump is off), which will vary from utility to utility.

Normally, this value is determined by hydraulic modeling of the system. In the absence of modeling data, pressure measured at a nearby hydrant during a time corresponding to peak water use can also be used.

No single correct value is available to measure pressure in the main. The problem is that the pressure in a main varies over the course of the day because of changes in water demand, pump operation, and tank water levels. Pressure also varies in the long term because of changes in pipe-carrying capacity, adjustment in pressure zone boundaries, and long-term changes in demands.

The designer sizing the service line should check with the utility concerning expected pressure at a tie-in point in the distribution system and any planned changes in expected pressure. Utilities should be careful when discussing pressures so as not to give the impression that they are guaranteeing a specific pressure under all conditions.

A conservative value for available pressure would be 20 psi since water utilities are generally minimally required to provide above 20 psi at the meter for fire protection. A nonconservative value would be a pressure reading taken at a nearby hydrant at a random time. A reasonable design value would be a pressure calculated by a model for peak flow conditions with nearby tank(s) at normal low level with an allowance for increased demands in the near future. Systems without a calibrated model must rely on pressure readings and use judgment on the amount to reduce pressure due to peak demands.

Another complicating factor is that the water use through the new service may itself affect pressure. A new residential tap on a 12-in. main will not affect pressure in the main. A 6-in. tap for an industrial customer on a 6-in. main, however, may affect the available pressure depending on the strength of the distribution system in that area and the potentially significant flow demand of the industrial customer. When the capability of the distribution system to serve the new customer is in doubt, a fire hydrant flow test may be appropriate to assess the distribution system capacity. When the piping is not yet installed and a utility has a calibrated hydraulic model of the remainder of the system, the model may also be used to estimate pressure under a variety of future conditions. When a new large-use customer is to be added (or old customer use greatly expanded), it is good practice to use the hydraulic model to try to maintain existing system conditions.

## Required Pressure

As with available pressure, there is no simple rule to determine the required pressure, or head, for the customer. Most customer fixtures will work over a wide range of pressures. At higher pressures, however, more flow will be delivered for a given opening of the faucet or valve.

Some fixtures will operate at pressures as low as 5 psi; however, the Universal Plumbing Code (UPC) and the National Standard Plumbing Code (NSPC) require a minimum 15 psi *residual* pressure (therefore, the customer service line pressure will need to be higher). The IPC requires a minimum 8 psi for faucets and 20 psi for pressure-balanced valves (some faucets require high pressures to meet the design flow rate). Some water-reducing fixtures today are requiring much higher pressures for performance that meets manufacturer specifications.

The height of fixtures above the service line also affects pressure. To provide 10 psi to a shower on the second story of a building, pressure of approximately 25 psi downstream of the meter in the basement is required. For most showers with pressure-balanced valves, the minimum pressure requirement is 20 psi.

As for the design flow rate, the required pressure for stand-alone piping for both domestic and fire supply for residential fire sprinkler system is prescribed by the 2019 NFPA 13D standard and the local plumbing code. The NFPA standard sets required flow and pressure requirements for one sprinkler head and, in most instances, for two sprinkler heads flowing simultaneously. Pressure delivered to each sprinkler head, under sprinkler flow conditions, is to be the higher of the minimum operating pressure given in the sprinkler listing or 7 psi. The design professional will need to consult both the plumbing code and NFPA 13D for pressure requirements in a multipurpose system.

The required pressure discussed in this chapter is the pressure needed downstream of meters and backflow preventers, not necessarily the pressure at customer fixtures. It is the responsibility of the plumbing designer to determine what pressure is required at that point. If the pressure is marginally below what is required, the service line, meter, or backflow preventer can be upsized to reduce head loss and raise pressure. However, if significantly more pressure is needed to provide service to upper floors of a building, pumping may be necessary. Pumping water from the distribution system adds an additional layer for analysis as significant withdrawal may lower distribution system pressures and create surge conditions that typically require further hydraulic analysis.

# HEAD LOSS

## Maximum Allowable Head Loss

The maximum allowable head loss through the service line, meter, and backflow preventer can be calculated as the difference between the head in the main and the head required by the customer. In US customary units, this is given by Eq. 5-1:

$$h_a = (z\ main - z\ cust) + 2.31\ (p\ main - p\ cust) \qquad \text{(Eq. 5-1)}$$

Where:

$h_a$ = maximum allowable head loss, *ft*
$z\ main$ = elevation of main, *ft*
$z\ cust$ = elevation corresponding to *p cust*, *ft*
$p\ main$ = pressure in main, *psi*
$p\ cust$ = pressure required by customer, *psi*
2.31 = conversion factor from psi to ft of water

It is not necessary to know the exact values of *z main* and *z cust* if one knows their difference (*z main – z cust).*

If $h_a$ is negative or very small (e.g., <5 ft), it may be possible to reevaluate and adjust some of the other parameters on the right side of the equation. Otherwise, if readjustment is not feasible, it may be necessary for the customer to install a booster pump to obtain design flow at a satisfactory pressure, being sure to maintain a suction pressure upstream of the pump of at least 5 psi and meet the required net positive suction head for the pump. In general, pumping is only acceptable for large customers. Most utilities do not allow individual residential customers to install booster pumps. The requirement to provide 5 psi of pump suction within the customer plumbing does not supersede the requirement to maintain at least 20 psi (or more in some jurisdictions) in distribution mains.

## Head Loss

It is necessary to use energy to move water from one point in the pipeline to another. The energy consumed cannot be recovered and is therefore "lost." In water distribution, this energy is usually reported in head (length) units, and the loss is referred to as *head loss.*

The head loss at the design flow rate must next be calculated for the piping configuration to determine if that flow can be delivered. If the available head exceeds the calculated head loss, that flow can be delivered. Methods for determining head loss for pipes, valves, and meters are discussed in the following sections.

## Piping Losses

When the length of a service line is substantial, pipe head loss is the most significant component of head loss. The pipe head loss can be calculated using many different equations, including the Hazen–Williams, Darcy–Weisbach, and Manning equations. In the United States, the Hazen–Williams equation is usually used. The Hazen–Williams equation for pipe head loss is written in terms of velocity. In US customary units, this is given as Eq. 5-2:

$$V = 1.318Cd^{0.63}\left(\frac{h}{L}\right)^{0.54} \qquad \text{(Eq. 5-2)}$$

Where:
- $V$ = velocity, *ft/s*
- $C$ = Hazen–Williams $C$ factor (unitless parameter)
- $d$ = diameter, *ft*
- $h$ = head loss, *ft*
- $L$ = length, *ft*

In most problems using English units, the diameter should be converted to inches and velocity converted to flow (velocity multiplied by cross-sectional area) in gpm. In this case, Eq. 5-2 becomes Eq. 5-3:

$$h = \frac{4.73L}{C^{1.852}D^{4.87}}Q^{1.852} \qquad \text{(Eq. 5-3)}$$

Where:
- $Q$ = flow rate, *gpm*
- $D$ = pipe inside diameter, *in.*

The Hazen–Williams $C$ factor depends on the piping material. Values of approximately 150 have been reported for plastic and copper tubing. Usually, a value of 130 is used as a conservative value for new pipe design.

For the small-size pipes used in service lines, the actual inside diameter (ID) of the pipe should be used and not the nominal diameter, because these two values can be significantly different depending on the class of pipe. Table 5-1 gives the typical ID corresponding to the nominal diameter for a variety of pipe materials and thicknesses (pressure class). Note that cross-linked polyethylene (PEX) tubing closely follows the sizing and head losses of polyethylene (PE) pipe. Commonly used PEX is SDR9 (see PE piping, per copper tubing sizing, IDR9).

Table 5-1 Actual internal diameter versus nominal diameter

| | Nominal diameter, in. | | | | | | | | | | |
|---|---|---|---|---|---|---|---|---|---|---|---|
| | ¾ | 1 | 1½ | 2 | 2½ | 3 | 3½ | 4 | 5 | 6 | 8 |
| **Material and Class** | **Actual internal diameter, in.** | | | | | | | | | | |
| Copper | | | | | | | | | | | |
| Type K | 0.745 | 0.995 | 1.481 | 1.959 | 2.435 | 2.907 | 3.385 | 3.857 | 4.805 | 5.741 | 7.583 |
| Type L | 0.785 | 1.025 | 1.505 | 1.985 | 2.465 | 2.945 | 3.425 | 3.905 | 4.875 | 5.485 | 7.725 |
| Type M | 0.811 | 1.055 | 1.527 | 2.009 | 2.495 | 2.981 | 3.459 | 3.935 | 4.907 | 5.881 | 7.785 |
| Ductile iron | | | | | | | | | | | |
| Class 50 | | | | | | | | | | 6.400 | 8.510 |
| Class 51 | | | | | | 3.460 | | 4.280 | | 6.340 | 8.450 |
| Class 52 | | | | | | 3.400 | | 4.220 | | 6.280 | 8.390 |
| Class 53 | | | | | | 3.340 | | 4.160 | | 6.220 | 8.330 |
| Polyethylene | | | | | | | | | | | |
| IDR9 | 0.681 | 0.875 | 1.263 | 1.653 | | | | | | | |
| PVC C900 | | | | | | | | | | | |
| Class 165 (DR25) | | | | | | | | 4.416 | | 6.282 | 8.240 |
| Class 235 (DR18) | | | | | | | | 4.266 | | 6.042 | 7.924 |
| Class 305 (DR14) | | | | | | | | 4.114 | | 5.796 | 7.602 |
| PVC ASTM D2241 | | | | | | | | | | | |
| Class 100 (SDR41) | | | | | | | | | | | |
| Class 125 (SDR32.4) | | | | | | | | | | | |
| Class 160 (SDR26) | | 1.155 | 1.714 | 2.153 | 2.615 | 3.190 | | 5.081 | | | |
| Class 200 (SDR21) | 0.890 | 1.149 | 1.680 | 2.109 | 2.561 | 3.126 | | 4.969 | | | |
| Class 250 (SDR17) | 0.886 | 1.121 | 1.636 | 2.055 | 2.497 | 3.038 | | 4.831 | | | |
| Steel | | | | | | | | | | | |
| Schedule 40 | 0.824 | 1.049 | 1.610 | 2.067 | 2.469 | 3.068 | 3.548 | 4.026 | 5.047 | 6.065 | 7.981 |

For those who prefer not to calculate the head loss, tables are provided in Appendix B to make it easy to determine head loss. Simply look up the head loss per unit length (typically per 100 ft) and multiply appropriately by the length of the pipe.

For a reasonable initial estimate of diameter on which to base sizing calculations for service lines, the diameter that results in a velocity of 5–8 ft/s at peak design flow can be used and calculate diameter as Eq. 5-4.

$$D = \sqrt{\frac{Q}{2.44V}} \qquad \text{(Eq. 5-4)}$$

The calculated diameter is rounded up to the next commercially available size. This value may need to be adjusted depending on the head available, number of appurtenances (e.g., meters, backflow preventers) in the line, and length of the line.

## Head Loss for Valves and Fittings

Numerous types of valves and fittings are used in service lines. Some typical valves are described in this section. Methods for determining head loss through the valves and fittings are presented in the following section. (General standards for service line valves and fittings are given in ANSI/AWWA C800. The manufacturer's specifications should also be checked.)

**Corporation valves or stopcocks.** These valves are used at the service tap in the water main and are usually left fully open. They are full-bore valves that cause minimal head loss.

**Corporation bends, gooseneck fittings.** For smaller pipe diameters, the pipe may be tapped at or near the top of the pipe, and there is a curved section of pipe called a corporation bend or gooseneck, which changes the direction of flow to horizontal heading toward the customer. The extra head loss because of this bend is usually relatively small compared to other minor losses.

**Curb stops.** These valves are used to start or stop customer service. They are usually full-bore ball valves with a minimal head-loss value located near the property line. An important consideration with curb stops and corporation stops is that they are normally fully opened. If they are left partly opened, they can cause significant head loss. For most utilities, the curb stop marks the end of the utility's piping and the beginning of the customer's piping.

**Meter resetters, yokes.** Meter yokes are a manufactured bracket that allows for placement of the meter. Resetters are used in meter pits that often adjust the height of the meter for easy access. Both resetters and yokes commonly have a shutoff valve on the main side of the meter. Appropriate head loss values or head loss across the yoke or resetter needs to be added into the head-loss calculations.

**Backflow prevention assemblies and check valves.** These specialty valves are installed to prevent backflow from the customer's plumbing into the distribution system. Their use is often mandated by building codes or state environmental regulations for buildings with a high risk of possible contamination should a backflow or back-siphonage condition occur in the distribution system. Guidance on the need and use of these devices is further described in M14 (AWWA 2015). This group includes a variety of assemblies from simple flap check valves to fully testable reduced-pressure backflow prevention (RPBP) assemblies. These valves usually have a great deal of head loss associated with them.

**Globe, angle, and gate valves.** These valves are used to isolate the customer's system. They are usually intended to be fully opened or fully closed. Throttling these valves will affect the ability of the system to deliver the design flow.

**Pressure-reducing valves.** These valves are used to protect the customer's plumbing from excessive pressure in the distribution system. They are usually placed between the curb stop and the meter and are owned by the customer. In most cases, at design flow rate, these valves may be fully opened because at such flow there will usually be sufficient head loss in the piping and meters that throttling the pressure is not needed. The head loss value will depend on the type of valve body (e.g., globe and angle).

## Minor Losses

*Minor losses* refer to head losses other than through the pipe itself and includes losses caused by valves, meters, backflow prevention assemblies, bends, and changes in diameter. These losses are usually a function of the velocity (and hence flow) squared. A problem in calculating these losses is that the data are usually given in a variety of forms including:

- Minor loss, $k$
- Minor loss, $Cv$
- Curve of minor loss versus flow rate
- Equivalent pipe length
- Orifice, $K$

Methods for calculating minor head loss ($h$) given each form of data are described in the following section.

**Minor loss, *k*.** The minor loss, *k*, is a dimensionless number that is used to relate head loss to velocity in Eq. 5-5:

$$h = \frac{kV^2}{2g} \qquad \text{(Eq. 5-5)}$$

Where:
$k$ = minor loss
$V$ = velocity, *ft/s*
g = acceleration due to gravity, *ft/s*$^2$

Substituting flow rate for velocity in the previous equation gives Eq. 5-6:

$$h = \frac{kQ^2}{383D^4} \qquad \text{(Eq. 5-6)}$$

Where:
$Q$ = flow rate, *gpm*
$D$ = diameter, *in*

Where the head loss is desired in psi ($p$), simply divide the head loss in feet by 2.31 to obtain the following:

$$p = \frac{h}{2.31} \qquad \text{(Eq. 5-7)}$$

Typical *k* values for valves, bends, and fittings are available from several sources (Crane 1969; Hydraulic Institute 1978; Walski 1994) and are summarized in Table 5-2. Minor losses for meters are given in Table 5-3, though it may be prudent to check graphs available from manufacturers. The meter values are based on the maximum allowable head loss at design flow rate according to ANSI/AWWA meter standards C700, C701, C702, C703, C704, C708, C710, C712, C713, C714, C715, and C750. Actual head-loss values are usually less than those stated in the standards. They may be only slightly less for meters with strainers but are often significantly less for meters without strainers. If strainers are doing a meaningful job of capturing material, there will be additional line loss. Individual meter manufacturers' curves should be checked.

**Table 5-2** Minor loss *k*-values

| Component | k-values |
|---|---|
| Gate Valve | |
| ¾ closed | 24 |
| ½ closed | 5.6 |
| ¼ closed | 1.2 |
| Fully open | 0.2 |
| Angle valve open | 2.5 |
| Globe valve | 6 |
| Swing check | 2 |
| Angle valve | 2.2 |
| Ball check | 4.5 |

*(continued)*

**Table 5-2** Minor loss *k*-values (*Continued*)

| Component | k-values |
|---|---|
| Butterfly valve | |
| Full open | 1.2 |
| 20° Closed | 1.4 |
| 40° Closed | 10 |
| 60° Closed | 94 |
| 80° Closed | 1,750 |
| Foot valve hinged | 2.2 |
| Foot valve pop | 12 |
| Basket strainer | 1 |
| 90° elbow | 0.3 |
| 90° long elbow | 0.15 |
| T - flow through run | 0.15 |
| T - flow through branch | 0.8 |
| Gooseneck | 3 |
| 90° bend | |
| r/d = 1* | 0.5 |
| r/d > 5 | 0.15 |
| 45° bend* | |
| r/d = 1 | 0.15 |
| r/d > 5 | 0.08 |
| Miter bends | |
| 90° | 1.2 |
| 60° | 0.5 |
| 30° | 0.14 |
| Expansion† | |
| d/D = 0.75 | 0.18 |
| d/D = 0.5 | 0.55 |
| d/D = 0.25 | 0.88 |
| Contraction† | |
| d/D = 0.25 | 0.43 |
| d/D = 0.50 | 0.33 |
| d/D = 0.75 | 0.18 |
| Entrance | |
| Projecting | 0.78 |
| Flush | 0.5 |
| Slightly rounded | 0.23 |
| Well rounded | 0.04 |
| Exit | 1 |

* r = bend radius; d = pipe diameter.

† Use velocity in smaller pipe for expansions and contractions. The word "reducer" is often used to cover both expansion and contraction, but direction of flow matters for head loss calculations.

**Table 5-3** Minor loss through meters

| Meter Type | Meter Size, *in.* | Safe Maximum Operating Capacity, *gpm* | Maximum Pressure Loss at Safe Maximum Operating Capacity, *psi* | *k* |
|---|---|---|---|---|
| Positive Displacement | ½ | 15 | 15 | 9.0 |
| | ½ × ¾ | 15 | 15 | 9.0 |
| | ⅝ | 20 | 15 | 5.1 |
| | ¾ | 30 | 15 | 4.7 |
| | 1 | 50 | 15 | 4.5 |
| | 1½ | 100 | 15 | 6.7 |
| | 2 | 160 | 15 | 8.3 |
| Turbine (Class I) | ¾ | 30 | 15 | 4.7 |
| | 1 | 50 | 15 | 5.3 |
| | 1½ | 100 | 15 | 6.7 |
| | 2 | 160 | 15 | 8.3 |
| | 3 | 350 | 15 | 8.8 |
| | 4 | 630 | 15 | 8.6 |
| | 6 | 1,300 | 15 | 10.2 |
| Turbine (Class II) | 1½ | 120 | 7 | 2.2 |
| | 2 | 190 | 7 | 2.7 |
| | 3 | 435 | 7 | 2.7 |
| | 4 | 750 | 7 | 2.8 |
| | 6 | 1,600 | 7 | 3.1 |
| | 8 | 2,800 | 7 | 3.2 |
| | 10 | 4,200 | 7 | 3.5 |
| | 12 | 5,300 | 7 | 4.6 |
| | 16 | 7,800 | 7 | 6.7 |
| | 20 | 12,000 | 7 | 6.9 |
| Fire service (Type II compound) | 3 | 350 | 12 | 7.0 |
| | 4 | 700 | 12 | 5.5 |
| | 6 | 1,600 | 12 | 5.4 |
| | 8 | 2,800 | 12 | 5.5 |
| | 10 | 4,400 | 12 | 5.5 |
| Fire service (Type III turbine with strainer) | 3 | 350 | 11 | 6.4 |
| | 4 | 700 | 11 | 5.1 |
| | 6 | 1,600 | 11 | 4.9 |
| | 8 | 2,800 | 11 | 5.1 |
| | 10 | 4,400 | 11 | 5.0 |

*(continued)*

**Table 5-3** Minor loss through meters (*Continued*)

| Meter Type | Meter Size, *in.* | Safe Maximum Operating Capacity, *gpm* | Maximum Pressure Loss at Safe Maximum Operating Capacity, *psi* | *k* |
|---|---|---|---|---|
| Propeller (main line) | 2 | 120 | 5 | 4.9 |
| | 3 | 300 | 5 | 4.0 |
| | 4 | 600 | 2 | 1.3 |
| | 5 | 900 | 1 | 0.7 |
| | 6 | 1,350 | 1 | 0.6 |
| | 8 | 1,800 | 0.5 | 0.6 |
| | 10 | 2,400 | 0.5 | 0.8 |
| | 12 | 3,375 | 0.5 | 0.8 |
| | 14 | 4,500 | 0.5 | 0.8 |
| | 16 | 5,700 | 0.5 | 0.9 |
| | 18 | 6,750 | 0.25 | 0.5 |
| | 20 | 8,250 | 0.25 | 0.5 |
| | 24 | 12,000 | 0.25 | 0.5 |
| | 30 | 18,000 | 0.25 | 0.6 |
| | 36 | 24,000 | 0.25 | 0.6 |
| | 42 | 40,000 | 0.1 | 0.2 |
| | 48 | 50,000 | 0.1 | 0.2 |
| | 54 | 55,000 | 0.1 | 0.2 |
| | 60 | 80,000 | 0.1 | 0.2 |
| | 66 | 95,000 | 0.1 | 0.2 |
| | 72 | 115,000 | 0.1 | 0.2 |
| Multijet | ⅝ | 20 | 15 | 5.1 |
| | ⅝ × ¾ | 20 | 15 | 5.1 |
| | ¾ | 30 | 15 | 4.7 |
| | 1 | 50 | 15 | 5.3 |
| | 1½ | 100 | 15 | 6.7 |
| | 2 | 160 | 15 | 8.3 |
| Single jet | ⅝ | 20 | 15 | 5.1 |
| | ⅝ × ¾ | 20 | 15 | 5.1 |
| | ¾ | 30 | 15 | 4.7 |
| | 1 | 50 | 15 | 5.3 |
| | 1½ | 100 | 15 | 6.7 |
| | 2 | 160 | 15 | 8.3 |
| | 3 | 320 | 15 | 10.5 |
| | 4 | 500 | 15 | 13.6 |
| | 6 | 1,000 | 15 | 17.2 |

*(continued)*

**Table 5-3** Minor loss through meters (*Continued*)

| Meter Type | Meter Size, *in.* | Safe Maximum Operating Capacity, *gpm* | Maximum Pressure Loss at Safe Maximum Operating Capacity, *psi* | *k* |
|---|---|---|---|---|
| Fluidic oscillator | ½ | 15 | 15 | 9.0 |
| | ½ × ¾ | 15 | 15 | 9.0 |
| | ⅝ | 20 | 15 | 5.1 |
| | ⅝ × ¾ | 20 | 15 | 5.1 |
| | ¾ | 30 | 15 | 4.7 |
| | 1 | 50 | 15 | 5.3 |
| | 1½ | 100 | 15 | 6.7 |
| | 2 | 160 | 15 | 8.3 |
| Residential fire sprinkler (meter only) | ¾ | 30 | 10.1 | 3.1 |
| | 1 | 50 | 10.7 | 3.8 |
| | 1½ | 100 | 7.7 | 3.4 |
| | 2 | 160 | 7.7 | 4.3 |
| Residential fire sprinkler (meter with strainer) | ¾ | 30 | 14.5 | 4.5 |
| | 1 | 50 | 15.3 | 5.4 |
| | 1½ | 100 | 11 | 4.9 |
| | 2 | 160 | 11 | 6.1 |
| Electromagnetic and Ultrasonic | | | | |
| Type 1 | ½ | 15 | 10 | 2.5 |
| | ⅝ | 20 | 10 | 3.4 |
| | ¾ | 30 | 10 | 3.1 |
| | 1 | 50 | 10 | 3.5 |
| | 1½ | 100 | 10 | 4.5 |
| | 2 | 160 | 10 | 5.5 |
| | 3 | 350 | 8 | 4.7 |
| | 4 | 600 | 8 | 5.0 |
| | 6 | 1,350 | 8 | 5.0 |
| | 8 | 1,600 | 8 | 11.3 |
| Type II | | | | |
| | ½ | 15 | 10 | 2.5 |
| | ⅝ | 20 | 10 | 3.4 |
| | ¾ | 30 | 10 | 3.1 |
| | 1 | 50 | 10 | 3.5 |
| | 1½ | 150 | 7 | 1.4 |
| | 2 | 220 | 7 | 2.0 |
| | 3 | 500 | 7 | 2.0 |
| | 4 | 880 | 7 | 2.0 |
| | 6 | 1,400 | 7 | 4.1 |
| | 8 | 2,800 | 7 | 3.2 |
| | 10 | 4,500 | 7 | 3.1 |
| | 12 | 5,500 | 7 | 4.2 |
| | 16 | 8,000 | 7 | 6.3 |
| | 20 | 12,000 | 7 | 6.9 |

**Minor loss, $C_v$.** For certain types of valves, minor losses are usually presented in terms of a carrying capacity instead of a *k*-value such that:

$$h = 2.31\left(\frac{Q}{C_v}\right)^2 \qquad \text{(Eq. 5-8)}$$

Where:

$C_v$ = minor loss carrying capacity, $\frac{gpm}{\sqrt{psi}}$

Unlike *k*, which is dimensionless and independent of diameter, $C_v$ has units and is a function of diameter. Manufacturers of valves usually provide a table giving the $C_v$ for a valve as a function of diameter. When $C_v$ is known at one diameter ($C_{v1}$) and it is desired to know it at a second diameter ($C_{v2}$), it can be approximated using Eq. 5-9:

$$C_{v_2} = C_{v_1}\left(\frac{D_2}{D_1}\right)^2 \qquad \text{(Eq. 5-9)}$$

Substituting back into the equation for head loss, the head loss can be approximated at other diameters ($D_2$), given the head loss at a single diameter ($D_1$) using the following equation:

$$h = 2.31\left(\frac{Q}{C_{v_1}}\right)^2\left(\frac{D_1}{D_2}\right)^4 \qquad \text{(Eq. 5-10)}$$

For computer models, it may be necessary to enter a minor loss, *k*. The *k* value can be calculated from $C_v$ at a given diameter using the following:

$$k = 885\frac{D^4}{C_v} \qquad \text{(Eq. 5-11)}$$

**Minor loss versus *Q* curve.** For some devices such as meters and backflow prevention assemblies, the head loss is given as a curve of head loss (or pressure drop) versus flow rate. For manual calculations, it is acceptable to simply look up the head loss from such a graph. For spreadsheet calculations, it may be necessary to determine an equation relating head loss and flow rate. The two types of curves are shown in Figure 5-1 and may be characterized as

- those that pass through the origin, which is typical of meters and most valves, and
- those that have a discrete head loss before they open, which is typical of backflow prevention assemblies.

For devices with head loss curves that pass through the origin, the head loss can be approximated as Eq. 5-12:

$$h = a\,Q^b \qquad \text{(Eq. 5-12)}$$

Where:

*a*, *b* = regression coefficients

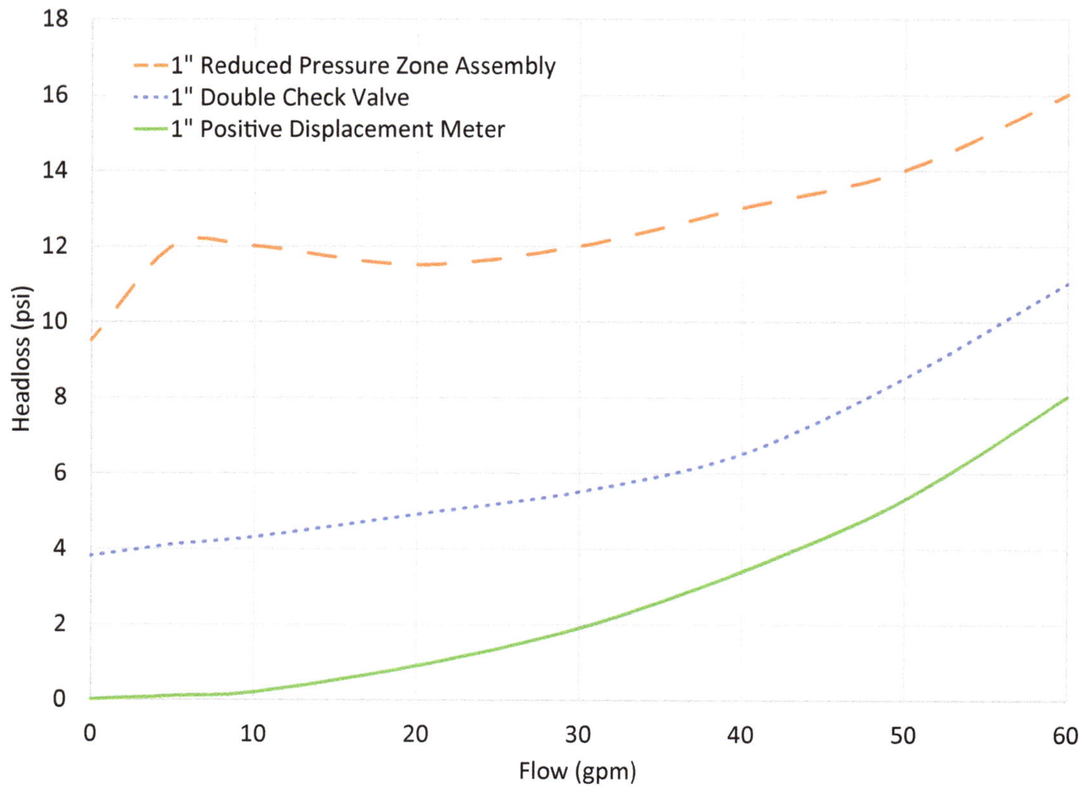


**Figure 5-1** Head loss versus flow rate curves

For most devices, $b$ is on the order of 2. The values for $a$ and $b$ can be found by using a regression program or by picking two points from the curve (in the range of the expected flows), such as $(Q_1,h_1)$ and $(Q_2,h_2)$, and calculating as follows:

$$b = \frac{\log\left(\frac{h_2}{h_1}\right)}{\log\left(\frac{Q_2}{Q_1}\right)}$$

$$a = \frac{h_1}{Q_1^b}$$

One problem with determining values of $a$ and $b$ is that they are a function of diameter. An alternative approach is to calculate the minor loss, $k$, for the meter. This method is usually not as good a fit to the manufacturer's curve, but it is relatively independent of diameter for a given meter or valve model. The value of $k$ can be determined by selecting a point near the design flow at the expected diameter and solving for $k$ using Eq. 5-13:

$$k = 885\frac{pD^4}{Q^2} \qquad \text{(Eq. 5-13)}$$

Where:

$p$ = pressure drop, *psi*
$Q$ = flow rate, *gpm*
$D$ = diameter, *in.*

When there is a discrete head loss before the valve opens (e.g., certain backflow prevention assemblies), the head loss can be given as Eq. 5-14:

$$h = h_{min} + aQ^b \qquad \text{(Eq. 5-14)}$$

Where:

$h_{min}$ = minimum head loss when flow, $Q = 0$, *ft*
$a, b$ = regression constants (differs from $a$ and $b$ of Eq. 5-12)

The values of $a$ and b can be calculated using the following equations:

$$b = \frac{\log\left(\frac{[h_2 - h_{min}]}{[h_1 - h_{min}]}\right)}{\log\left(\frac{Q_2}{Q_1}\right)} \qquad \text{(Eq. 5-15)}$$

$$a = \frac{h_1 - h_{min}}{Q^b} \qquad \text{(Eq. 5-16)}$$

As was the case with valves, the curve of which passes through the origin, it is possible to model the loss using a minor loss, $k$. The modeling can be done by selecting a point from the curve at the design flow expected and calculating $k$ as Eq. 5-17:

$$k = 885\frac{(p - p_{min})D^4}{Q^2} \qquad \text{(Eq. 5-17)}$$

Where:

$$P\,min = \frac{h_{min}}{2.31}, psi$$

Valves with a discrete pressure drop are handled differently in different pipe network hydraulic models. In some cases, they can be modeled as a negative pump curve; in others, they can be modeled as a discrete pressure drop in series with a minor loss; and in still others, they can be described directly to the model as a series of data points on a head loss versus flow curve. The $k$ values for backflow prevention assemblies as given in Table 5-4 are based on the values given in the ANSI/AWWA standards C510 and C511 and, as is the case with meters, values for actual assemblies are usually less.

**Table 5-4** Minor loss through backflow prevention assemblies

| Type | Nominal Size, *in.* | Max Q, *gpm* | P open, *psi* | Max P *psi* | *k* |
|---|---|---|---|---|---|
| Reduced pressure | ½ | 7.5 | 10 | 22 | 4.6 |
| | ¾ | 30 | 10 | 20 | 3.1 |
| | 1 | 50 | 10 | 18 | 2.8 |
| | 1¼ | 75 | 10 | 18 | 3.1 |
| | 1½ | 100 | 10 | 16 | 2.7 |
| | 2 | 160 | 10 | 16 | 3.3 |
| | 2½ | 225 | 10 | 16 | 4.1 |
| | 3 | 320 | 10 | 15 | 3.5 |
| | 4 | 500 | 10 | 14 | 3.6 |
| | 6 | 1,000 | 10 | 14 | 4.6 |
| | 8 | 1,600 | 10 | 14 | 5.7 |
| | 10 | 2,300 | 10 | 14 | 6.7 |
| Double check valve | ½ | 7.5 | 4 | 10 | 2.3 |
| | ¾ | 30 | 4 | 10 | 1.9 |
| | 1 | 50 | 4 | 10 | 2.1 |
| | 1¼ | 75 | 4 | 10 | 2.3 |
| | 1½ | 100 | 4 | 10 | 2.7 |
| | 2 | 160 | 4 | 10 | 3.3 |
| | 2½ | 225 | 4 | 10 | 4.1 |
| | 3 | 320 | 4 | 10 | 4.2 |
| | 4 | 500 | 4 | 10 | 5.4 |
| | 6 | 1,000 | 4 | 10 | 6.9 |
| | 8 | 1,600 | 4 | 10 | 8.5 |
| | 10 | 2,300 | 4 | 10 | 10.0 |

Note: *k* based on $885 \times (P_{max} - P_{open}) \times \frac{D^4}{Q^2}$.

**Equivalent length.** In some cases, an equivalent pipe length (*L*/*D*) is specified instead of a minor loss, *k*. The minor loss, *k*, can be determined from the Hazen–Williams equation as:

$$k = 194C^{1.852}D^{0.167}V^{0.148}\left(\frac{L}{D}\right) \quad \text{(Eq. 5-18)}$$

where *C*, *D*, and *V* refer to the properties of the equivalent pipe. For most cases, this can be simplified to Eq. 5-19:

$$K = 0.02\left(\frac{L}{D}\right) \quad \text{(Eq. 5-19)}$$

and the minor loss, *k*, can be used as previously described.

**Orifices.** In most cases, it is not necessary to determine head loss through the actual orifice, nozzle, or sprinkler head, but there are some situations where it does become necessary. Most orifices can be given by an equation similar to the equation where $C_v$ is previously defined. For orifices discharging to the atmosphere, the downstream pressure is atmospheric.

In some applications, sprinkler manufacturers describe the head loss through their sprinklers using a *K* value, which should not be confused with the minor loss, *k*. The *k* value is a carrying-capacity value like $C_v$. Solving for flow rate through a sprinkler can best be done using a hydraulic model (especially if more than one sprinkler is flowing at a time) (Walski 1994).

For details on sizing services for fire sprinkler systems, the reader should consult NFPA 13 (NFPA 2019a) for commercial and industrial buildings; NFPA 13R (NFPA 2019b) for residential occupancies up to four stories; and NFPA 13D (NFPA 2019c) for residential structures and manufactured buildings of three stories or smaller. Readers can also consult *Residential Fire Sprinkler Systems: Guidance for Water Utilities* (AWWA 2018) and the *Water Purveyor's Guide to Fire Sprinklers in Single Family Dwellings* (National Fire Sprinkler Association 2006), which give information specific to water utilities in creating policies and requirements for water supply to sprinklered dwellings. These publications include information on the applicable codes and standards, residential fire meters, backflow prevention, installation arrangements, hydraulic requirements, fees and charges, inspection, testing and maintenance, and options for rural water systems. Another good resource is the guide *Understanding Water Supply for Home Fire Sprinkler Systems* (Home Fire Sprinkler Coalition 2012).

## PERFORMING SIZING CALCULATIONS

The approach used to select and validate the service size depends on the complexity of the system and the form in which the data are presented. Approaches range from simple manual calculations based on consultation with tables to sophisticated computerized hydraulic models of the service line and piping.

For single-family residences (without a residential fire sprinkler system or an extensive irrigation sprinkling system), a ¾-in. or 1-in. service line with a ⅝-in. meter is typical in North America. (Some systems may require larger service lines for customers with lawn sprinklers and pools.) In general, the cost to install a larger service line when the building is being constructed is very small, whereas the cost to upsize later is significant. It is best to be conservative when sizing service lines. For example, assume the pressure in the main is 80 psi, and the customer only needs 30 psi. The service line can be sized for a pressure drop of 50 psi, but for only a slightly greater cost the service line can be sized to provide a 10-psi drop and give the customer 70 psi. The decision is a trade-off between quality of service and cost. Where longer service lines cause significant head loss, utilities could eliminate many of their low-pressure complaints by conservatively sizing their service lines.

Using a standard pipe size for a residential customer (e.g., ¾ in. or 1 in.) may not be adequate if:

- the pressures in the main are marginal,
- the customer fixtures and appliances are located at a significantly higher elevation than the main,
- the customer is set back far from the main, or
- the customer has some large water uses (e.g., fire or irrigation sprinkler systems).

Even for single-family residences, some hydraulic analysis may be necessary, notably homes requiring residential fire sprinklers.

All the calculations start with the determination of the design flow, the available pressure, required pressure, length of the service line, backflow prevention requirements, and elevations of both the main and the customer's system. Flow and pressure requirements for residential fire sprinklers are provided in the NFPA 13D standard. Once the flow and pressure requirements are known, the designer then selects the size of the piping, valves, and meters and calculates the head loss at the design flow rate or the flow rate that can be delivered using the available head. Pipe and fixture sizes are adjusted until the flow and head requirements are met without excessive cost.

Backflow prevention requirements are often specified by the water supplier's cross-connection protection policy. Double check valve assemblies or reduced-pressure backflow assemblies are most commonly used. Double check valves have lower head loss; their head loss varies directly with flow, which provides opportunity to reduce losses by increasing the assembly size. Reduced-pressure backflows have higher head loss and, because of their inherent design, the loss is incurred even at low flows regardless of size. Only reduced-pressure backflow preventers are suitable for high-hazard applications.

Once the flow and pressure requirements are known, the designer then selects the size of the piping, valves, backflow preventers, and meters and calculates the head loss at the design flow rate or the flow rate that can be delivered using the available head. Pipe and fixture sizes are adjusted until the flow and head requirements are met without excessive cost.

In addition to sizing the service line to provide adequate pressure, the size should also be selected to prevent water hammer (surge). Velocities greater than 10 ft/s should be avoided, even if the service line can provide adequate pressure at these high velocities. The maximum recommended velocities for some materials may be substantially lower than 10 ft/s, and the frequency and duration of changes in velocity must be reviewed. The pipe manufacturer should be consulted to obtain the recommended maximum velocity. If surge is a problem, it can be mitigated through good engineering practice.

Figure 5-2 shows a schematic of head loss from the water main to the customer. Design involves selecting a large-enough pipe and meter such that the drop in the hydraulic grade line is acceptable. The steps for determining the size of pipes, meters, and other fittings are shown in the flow chart in Figure 5-3.

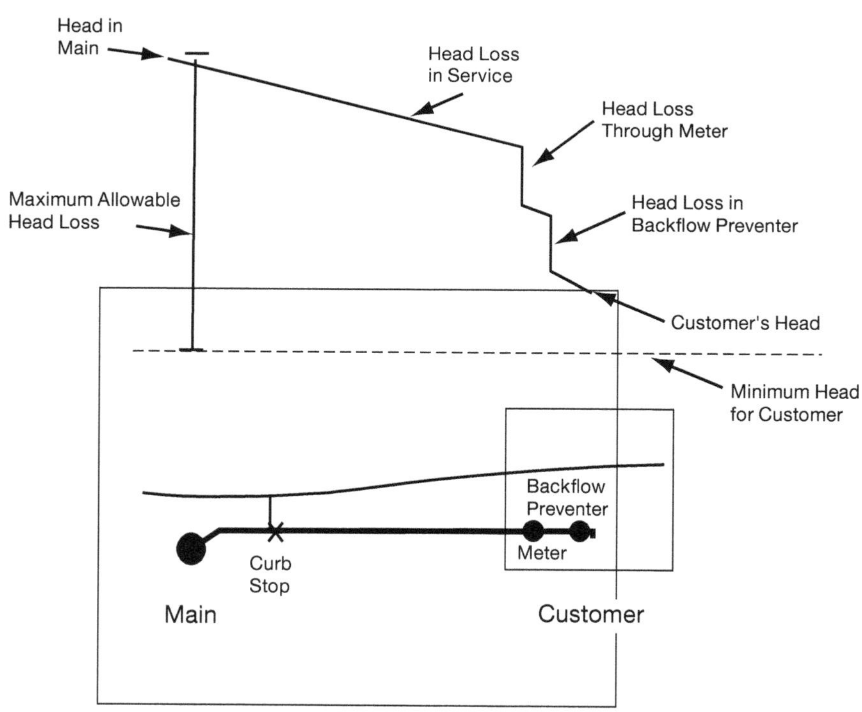


**Figure 5-2** Head loss from water main to customer

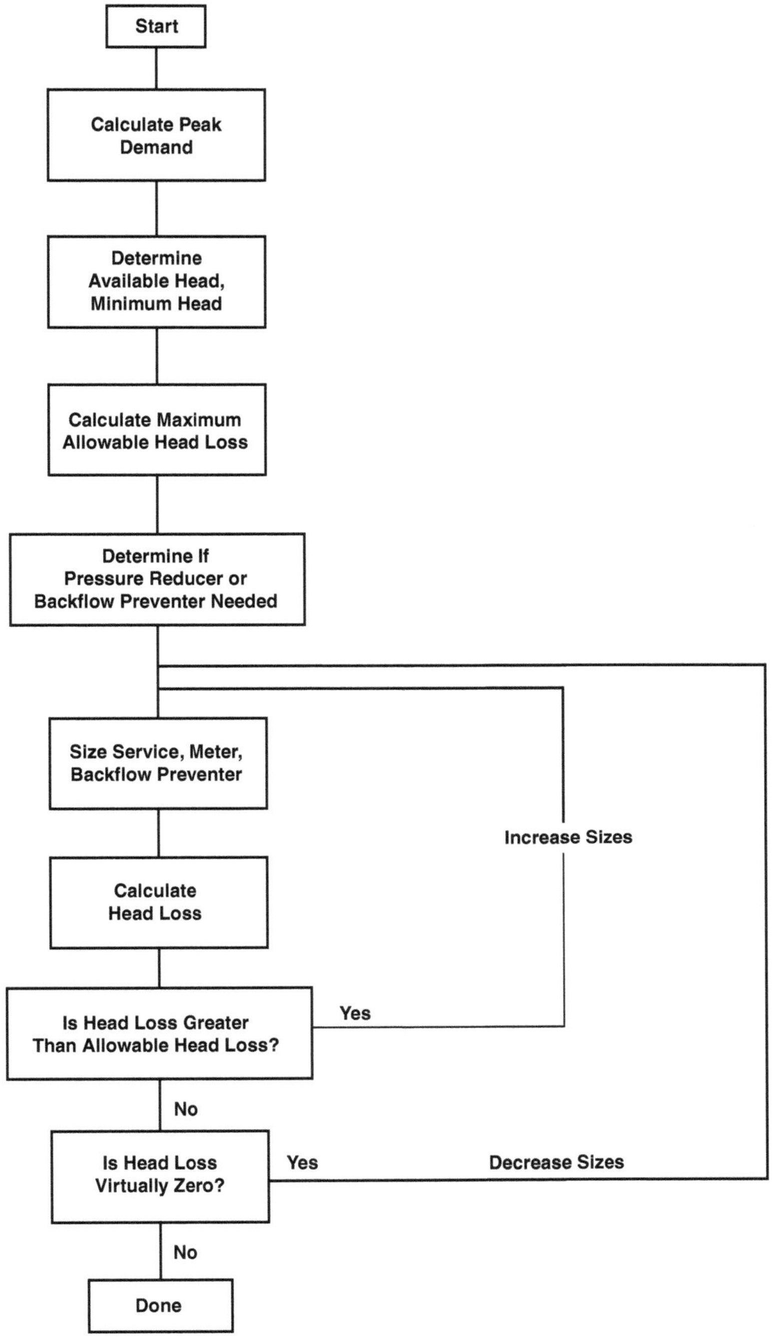


**Figure 5-3** Steps for determining pipe, meter, and other fitting sizes

At one extreme, the calculations can be performed by looking up the head loss through pipes and valves in tables as given in Appendix B. This approach does not require the use of a computer but is the most cumbersome when performing iterative calculations trading off pipe size and cost. To simplify the repetitive calculations, a spreadsheet can be set up to perform the calculations. Because there are similarities between customers, a template spreadsheet is suggested. The challenge is to convert the graphs of head loss (or pressure drop) versus flow, for meters and backflow valves provided by manufacturers, into equations using the methods described previously. Once the equations are derived, however, it is easy to perform repeated calculations.

When the service line is complicated with branches or if the designer wants to calculate the hydraulics of the system down to sprinkler heads and outlets, a hydraulic model is the ideal approach to sizing. In this case, the designer must enter the coefficients for each device into the model.

While most designers would initially feel that the manual approach is the easiest for a single calculation, spreadsheets and models are much easier should the customer come back with changes. For example, "Suppose we change the design flow to 120 gpm," or "We just found out we will need a minimum of 45 psi to make the equipment function." With a computerized calculation, repeating the calculation for these "what if" scenarios is easy.

Each of these approaches (manual calculations, spreadsheet calculations, and hydraulic models) is illustrated in the following figures by calculating a recommendation for the same problem.

## Example Sizing Problem

In this problem, there is a service line with the following characteristics:

| | | |
|---|---|---|
| Design flow rate | = | 75 gpm |
| Service length | = | 200 ft |
| Elevation of main | = | 585.2 ft |
| Elevation of meter | = | 598.1 ft |
| Design pressure in main | = | 45 psi |
| Required pressure for customer | = | 20 psi |
| Pipe material | = | Type L copper tubing |

There will be a compound meter and a RPBP assembly. The pressure drop versus head loss curves for these devices are shown in Figures 5-4 and 5-5. Losses through the corporation stopcock and curb stop can be considered negligible. Use a conservative value of *C* as 130.

$$\text{Available head loss} = 585.2 - 598.1 + 2.31\ (45 - 20) = 44.8\ \text{ft}$$

Manual, spreadsheet, and hydraulic model approaches to solving this problem are illustrated in the following sections.

**Manual calculation.** The manual approach is best suited for simple systems in which the range of choices in size is limited (e.g., typical domestic system with no sprinklers). In manual calculations, pressure drop in psi is sometimes used in place of head loss in feet. Convert the 45 ft available head to pressure drop in psi.

$$p = \frac{45}{2.31} = 19.5\ \text{psi allowable pressure drop}$$

Assume, initially, that a 2-in. nominal size pipe with a 2-in. meter and a 2-in. RPBP assembly will be used. The velocity in 2-in. copper tubing (actual ID = 1.985 in.) is 7.76 ft/s using the tables in Appendix B. For manual calculations, the entrance loss into the service line can be ignored. The pressure drop per 100 ft of piping is 5.8 psi. For 200 ft of line, this rate results in a pressure drop of 11.6 psi.

From Figure 5-4, the pressure drop in the meter is 1.5 psi. From Figure 5-5, the pressure drop through the RPBP assembly is 11.0 psi.

Summing the pressure drops gives the total pressure drop as:

$$p = 11.6 + 1.5 + 11.0 = 24.1\ \text{psi}$$

which exceeds the allowable pressure drop. The pipe, meter, and RPBP assembly need to be upsized and the calculations repeated until a satisfactory combination is found. Leaving the size of the RPBP assembly and meter the same and upsizing the pipe to 2½ in.

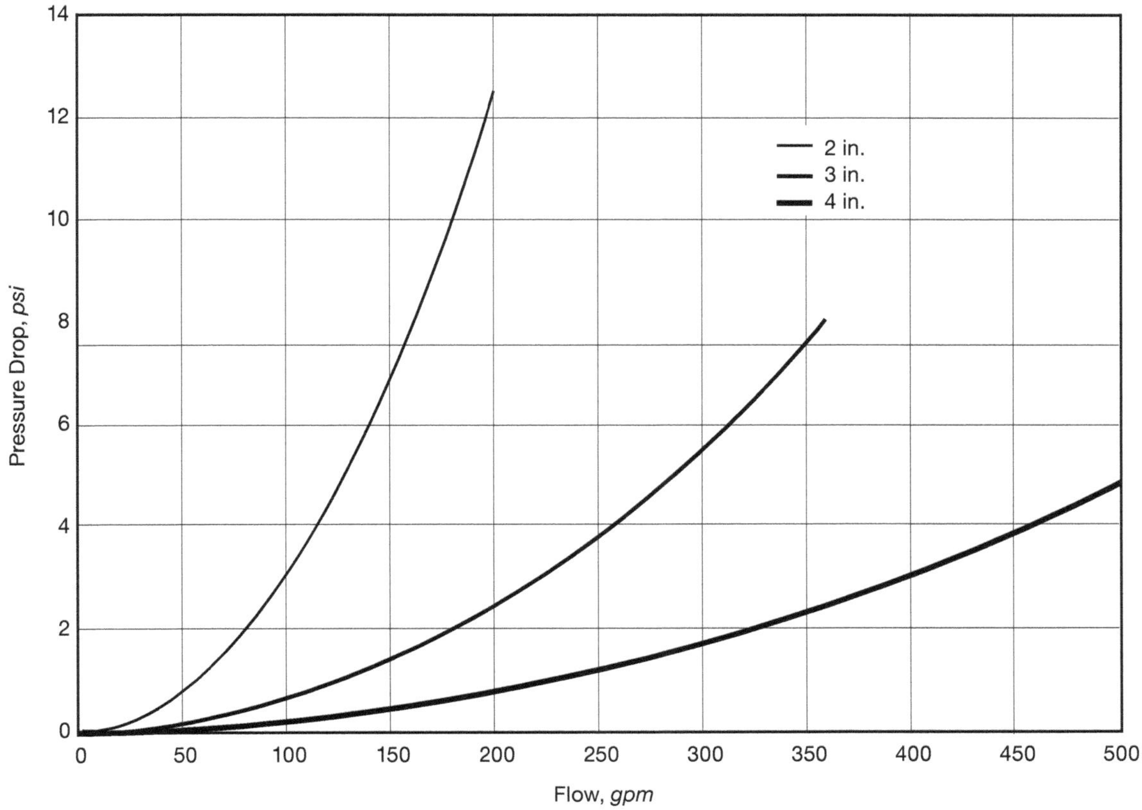


**Figure 5-4** Compound-meter pressure drop versus flow curves

will prove satisfactory in this instance. A larger pipe size may be desirable to provide better pressure and to provide a margin for error.

**Spreadsheet calculation.** It takes longer to set up a spreadsheet, but repeat calculations are easy. For the same example addressed by the manual calculation, the head loss equations used are summarized as follows:

Pipe loss:

Using Eq. 5-3:

$$k = (10.4)(200)\left(\frac{75}{130}\right)^{1.852} D^{-4.87} = 752D^{-4.87}$$

Entrance loss:

Using Eq. 5-6:

$$h = 0.5\frac{75^2}{383D^4}$$

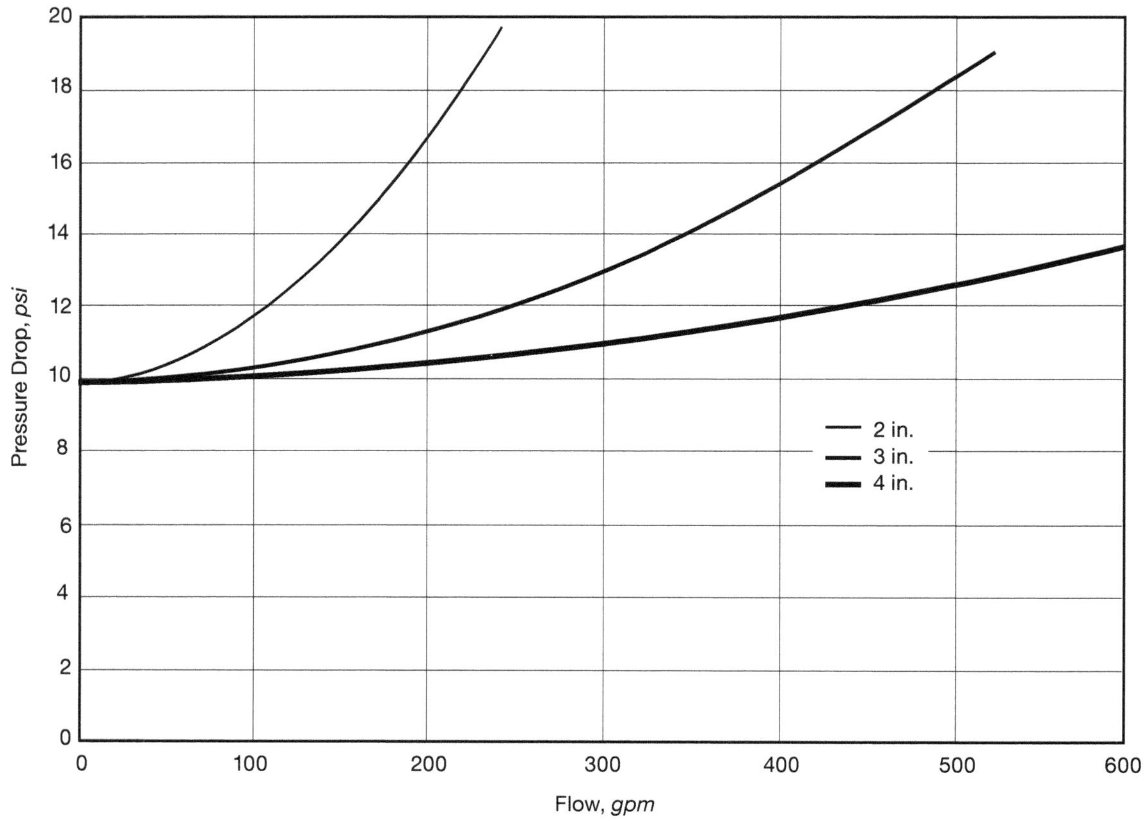


**Figure 5-5** RPBP assembly pressure drop versus flow curves

Meter loss: The meter in question has a pressure drop of 8 psi at 160 gpm in the 2-in. size. Therefore, from Eq. 5-13:

$$k = 885\frac{(8)(2^4)}{160^2} = 4.42$$

Head loss: The head loss can be calculated using Eq. 5-6 as:

$$h = 4.42\frac{75^2}{383D^4} = \frac{64.9}{D^4}$$

Backflow prevention assembly: The backflow-prevention assembly has a value for $h_{min}$ of 10 psi or 23.1 ft, and the $k$ value can be calculated using 14-psi drop in the 2 in. at 160 gpm as:

$$k = 885\frac{(p - p_{min})D^4)}{Q^2} = 2.2$$

For the 2-in. assembly, this gives:

$$h = 23.1 + 2.2\frac{75^2}{383D^4}$$

**Table 5-5** Service-line sizing calculation spreadsheet (L = 200 ft, Q = 75 gpm)

| Variables | Variation 1 | Variation 2 | Variation 3 | Variation 4 | Variation 5 | Variation 6 |
|---|---|---|---|---|---|---|
| *D*, in nominal pipe | 2 | 2.5 | 3 | 2.5 | 3 | 3.5 |
| *D*, in meter | 2 | 2.5 | 3 | 2 | 3 | 2 |
| *D*, in RPBP | 2 | 2.5 | 3 | 2 | 2.5 | 2.5 |
| *D*, in actual pipe | 1.985 | 2.465 | 2.945 | 2.465 | 2.465 | 3.425 |
| Available head, *ft* | 44.8 | 44.8 | 44.8 | 44.8 | 44.8 | 44.8 |
| **Calculated parameters** | | | | | | |
| Pipe loss, *ft* | 26.7 | 9.3 | 3.9 | 9.3 | 9.3 | 1.9 |
| Entrance loss, *ft* | 0.5 | 0.2 | 0.1 | 0.2 | 0.1 | 0.0 |
| Meter loss, *ft* | 4.1 | 1.7 | 0.8 | 4.1 | 0.8 | 4.1 |
| RPBP, loss | 25.3 | 24.0 | 23.5 | 25.3 | 24.0 | 24.0 |
| Total loss, *ft* | 56.5 | 35.1 | 28.3 | 38.8 | 34.2 | 30.0 |
| Available loss, *ft* | 11.7 | 9.7 | 16.5 | 6.0 | 10.6 | 14.8 |
| Velocity, *ft/s* | 7.8 | 5.1 | 3.5 | 5.1 | 3.5 | 2.6 |

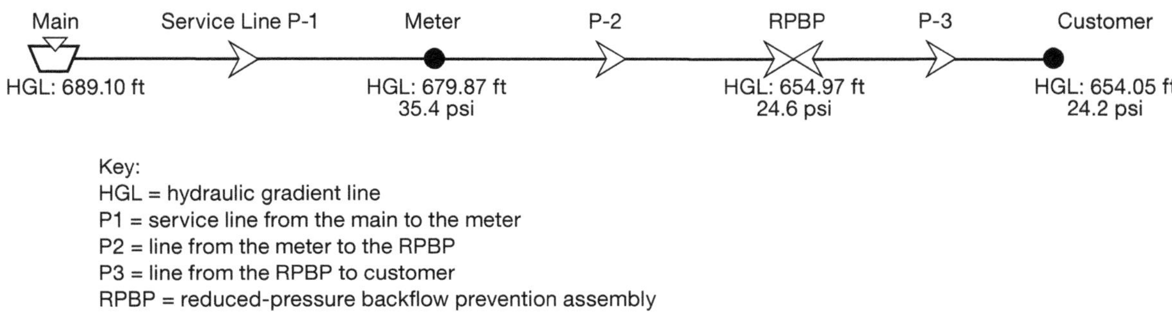


**Figure 5-6** Hydraulic model used to determine head loss: 75 gpm, 2.5-in. meter scenario

The results of the spreadsheet calculation are shown in Table 5-5, and they indicate that a 2.5-in. diameter service line and meter would be acceptable, as shown in Variation 2. Because this is an uncommon size, a 3-in. diameter is selected. Note that variations 3 through 6 would also be viable design alternatives.

**Hydraulic model.** The hydraulic model used to determine the head loss is shown in Figure 5-6. With a model, the service line can be represented as a single pipe with several minor losses corresponding to the service line or several pipes in series—e.g., the service line P1 from the main to the upstream side of the meter, the line P2 from the meter to the backflow prevention assembly, and the line P3 downstream from the backflow prevention assembly to the point where the customer pressure is measured and analyzed individually.

With a model, more information can be obtained than from a manual calculation, such as the pressure at any point along the line, the velocity in each pipe, and the hydraulic grade line (head) at any point along the line. Once a model is set up, the "what if " calculations are easy to perform.

**Selection.** For the sample problem addressed in the preceding discussion, several alternative combinations of pipe and valve sizes will provide adequate flow capacity for the head available. The final decision will be based on the economics of the situation, the accuracy range of the meter, and the amount of safety factor the designer would like to include.

## REFERENCES

American National Standards Institute/American Water Works Association (ANSI/AWWA). C510. Double Check Valve Backflow Prevention Assembly. Denver: AWWA.

ANSI/AWWA. C511. Reduced-Pressure Principle Backflow Prevention Assembly. Denver: AWWA.

ANSI/AWWA. C700. Cold-Water Meters—Displacement Type, Metal Alloy Main Case. Denver: AWWA.

ANSI/AWWA. C701. Cold-Water Meters—Turbine Type, for Customer Service. Denver: AWWA.

ANSI/AWWA. C702. Cold-Water Meters—Compound Type. Denver: AWWA.

ANSI/AWWA. C703. Cold-Water Meters—Fire-Service Type. Denver: AWWA.

ANSI/AWWA. C704. Propeller-Type Meters for Waterworks Applications. Denver: AWWA.

ANSI/AWWA. C708. Cold-Water Meters—Multijet Type. Denver: AWWA.

ANSI/AWWA. C710. Cold-Water Meters—Displacement Type, Plastic Main Case. Denver: AWWA.

ANSI/AWWA. C712. Cold-Water Meters—Singlejet Type. Denver: AWWA.

ANSI/AWWA. C713. Cold-Water Meters—Fluidic-Oscillator Type. Denver: AWWA.

ANSI/AWWA. C714. Cold-Water Meters for Residential Fire Sprinkler Systems in One- and Two-Family Dwellings and Manufactured Homes. Denver: AWWA.

ANSI/AWWA. C715. Cold-Water Meters—Electromagnetic and Ultrasonic Type, for Revenue Applications. Denver: AWWA.

ANSI/AWWA. C750. Transit-Time Flowmeters in Full Closed Conduits. Denver: AWWA.

ANSI/AWWA. C800. Underground Service Line Valves and Fittings. Denver: AWWA.

AWWA. 2015. M14, *Recommended Practice for Backflow Prevention and Cross-Connection Control.* Denver: AWWA.

AWWA. 2018. *Residential Fire Sprinkler Systems: Guidance for Water Utilities*, Denver: AWWA.

Brock, P.D. 1993. *Fire Protection Hydraulics and Water Supply Analysis*. Fire Protection Publications. Stillwater, Okla.: Oklahoma State University.

Crane. 1969. "Flow of Fluids through Valves, Fittings, and Pipe." Technical Paper 410. Stamford, Conn.: Crane Co.

Home Fire Sprinkler Coalition (HFSC). 2012. "Understanding Water Supply for Home Fire Sprinkler Systems." Frankfort, Ill.: HFSC.

Hydraulic Institute. 1978. *Engineering Data Book*. Cleveland, Ohio: Hydraulic Institute.

National Fire Protection Association (NFPA). 2019a. NFPA 13, Standard for the Installation of Sprinkler Systems. Quincy, Mass.: NFPA.

NFPA. 2019b. NFPA 13R, Standard for the Installation of Sprinkler Systems in Low-Rise Residential Occupancies. Quincy, Mass.: NFPA.

NFPA. 2019c. NFPA 13D, Standard for the Installation of Sprinkler Systems in One- and Two-Family Dwellings and Manufactured Homes. Quincy, Mass.: NFPA.

National Fire Sprinkler Association (NFSA). 2006. "Water Purveyor's Guide to Fire Sprinklers in Single Family Dwellings." Patterson, N.Y.: NFSA.

Walski, T.M. 1984. *Analysis of Water Distribution Systems*. New York: Van Nostrand Reinhold.

Walski, T.M. 1994. "An Approach for Handling Sprinklers, Hydrants, and Orifices in Water Distribution Models." Proc. AWWA ACE Convention, Anaheim, Calif.

Walski, T.M. and Clyne, R.E. 1996. "Sizing Residential Service Lines." *Journal AWWA*. 88(10):70–78.

Chapter 6

# Meter Sizing

## INTRODUCTION

The water meter is a changeable component of the customer's water system. Unlike the service line and water tap, which will require expensive excavation and retapping when incorrectly sized, water meters can usually be changed less expensively. The purpose of this chapter is to present guidelines on meter selection based on actual or likely water demands. Undersizing the meter can cause pressure-related problems; it can also result in a higher volume of apparent water loss with associated lost revenue at high-end flows and inaccurate meter recordings. Oversizing the meter means overspending for the meter—a higher volume of apparent water loss with associated reduced revenue at low-end flows because of inaccurate meter recordings.

## REVENUE IMPLICATIONS

The sizing of the meters and services can have significant impacts on a utility's revenue. As the flow drops below the recommended flow range for a meter, the meter will under-register. Oversized meters can result in lost revenue because of inaccurate registration at low flow. Oversized residential meters may fail to register toilet leaks, a common occurrence in household plumbing systems. Such leaks result in considerable wasted water, and accurate metering at low flows is necessary for detection. From the standpoint of registration accuracy and revenue, water meters should not be sized conservatively, i.e., larger than needed. In addition to charging for water consumed, many water utilities have a charge based on tap (or service line) size or meter size—either every billing period or at the time of installation (e.g., tapping fee or capital recovery fee). Sizing decisions may impact revenue through fixed charges based on meter size.

The recurring fixed charge is typically billed for a variety of routine services, such as customer meter reading, billing, and separate services such as stormwater management.

Numerous water utilities have begun to charge for stormwater services as a distinct land parcel-based component on the water bill, thus lowering the fixed charge based upon either line or meter size. The designer should stay abreast of changes in the customer billing rate structure that are based upon either line or meter size and understand the impacts that such changes might exert in meter sizing decisions.

To size a water meter appropriately, the designer needs to understand the utility's design criteria and determine the demand flow range of the customer's system. For existing services, empirical demand data may be the best source of information to ensure that the meter will meet actual needs. For new services, estimated demand must be calculated from other data. Chapters 3 and 4 provide appropriate methods for estimating design flows. The objective is to ensure that the meter is of sufficient size to properly handle the likely demands as estimated by the designer.

The designer needs to understand all aspects of the customer's demands. Relevant demand information may be obtained by having the customer complete a form that describes the anticipated nature of the flow (e.g., steady always, steady flow during working day, high peaks when filling tanks, high peaks during irrigation, etc.), peak demand, flow range, inside or outside meter setting, and, for utilities that require metering of fire sprinkler systems, data regarding fire service demands. Sprinkler and fire service demands covered in-depth in Chapter 4 are especially important considerations because they have a significant impact on meter type and size but do not appear on typical demand profiles based strictly on normal daily or weekly usage. This information will assist the designer in determining the size and type of meter (i.e., positive displacement, multijet, singlejet, electromagnetic, ultrasonic, compound, or turbine) that will most accurately register water demands and meet the customer's needs. Note that water usage patterns at an existing metered site may change significantly over time. For example, older factory sites may be repurposed with conversions to apartments, retail, or light commercial applications. When water use changes, the water meter size and/or type may also need to be changed.

## SYSTEM COMPONENTS POLICY REVIEW

After estimating the customer's peak demand and demand flow range as described in Chapters 3 and 4, the next step is to collect data regarding the system components. To complete this reconnaissance, the following field information may be required:

- flow tests of the water main that serves the area, if available;
- pressure in the service provider's water main during the customer's peak demand periods, if available;
- elevation differential of city water main and customer's indoor piping;
- length of service pipe necessary to extend from the water main to the meter;
- the proposed size of pipe used by the customer from the meter to the building;
- effective pipe distance for the service line's fittings and valves;
- utility's engineering standards for service line, appurtenances, and meter installation;
- requirements for Residential Fire Sprinkler Systems (RFSS) typically following the NFPA 13D Standard for the Installation of Sprinkler Systems in One- and Two-Family Dwellings and Manufactured Homes (NFPA 2019c); and
- fire-line meter requirements for systems typically following the NFPA 13 (NFPA 2019a) and 13R (NFPA 2019b) standards for buildings other than one- and two-family dwellings and manufactured homes.

Information regarding the system flow capability, pressure, and depth and location of the water main, along with fire-line meter requirements, may be available from the utility. Elevation differential, length and size of service pipe, and fitting and valve specifications are information that the designer will need to determine. After obtaining this information, the designer should proceed with obtaining information about the utility's engineering standards and policies so that the service line and meter are installed according to the utility's specifications. Examples of questions the designer may want to ask regarding the utility's standards and policies to determine the proper meter size include the following:

- Does the utility require the meter to be located at the property line or inside of the customer building?
- Does the utility allow the tapping of a separate fire line if one is required?
- Do the service line and meter need to be the same size?
- What are the utility's required service line and meter-setting design requirements?
- Does the utility require the fire sprinkler line (if one exists) to be metered?
- What are the service line specifications to and from the meter?
- Are there maximum head loss or velocity requirements?

## WATER METER SELECTION

Water meters are comparatively easy to replace; accordingly, if a meter is not properly sized, the expense of replacement is generally not cost prohibitive if sufficient space is available in the meter pit, vault, or the meter room inside a building.

The designer should refer to each water meter manufacturer's specifications to determine if the meter selected to serve the customer can (1) accurately record the predicted maximum flow for short periods of time without damage or without above-normal wear occurring to the meter and (2) accurately record the anticipated range of flows. AWWA has established standards for cold-water meters on flow range capabilities and maximum pressure losses that can be used by the designer. As new meter designs are developed that can operate accurately at different ranges of flow, the designer should check with the utility to determine if such new meters are in accordance with the utility's requirements.

The designer should use utility-approved meter manufacturers' specifications whenever possible to determine pressure loss, range of accuracy, and capacity. Table 6-1 provides minimum, normal, and maximum flow range values for water meters based on ANSI/AWWA Standards C700 through C710, and C712, C713, C714, and C715. These standards should be used as a guideline when meter manufacturers' data are not readily available. The designer must be aware that there are several types of meters, and that they have different capacities and ranges of accuracies. These tables are only a guide.

A meter's normal flow range is the range within which meter performance is optimal. For mechanical meters (e.g., positive displacement, multijet, singlejet, compound, and turbine), excessive flow above the high normal rate for sustained periods of time may cause excessive wear. Although electromagnetic and ultrasonic meters may have reduced accuracy during periods of flow above the high normal rate (Sumrak et al. 2016), these meter types are not permanently damaged by excessive flow above the high normal rate for sustained periods of time.

Concern for surge or deluge flows in ⅝-in. × ¾-in. water meters—the traditional residential service meter—is important when sizing water meters for residential fire sprinkler systems (RFSS), where the design flow usually exceeds the rated

**Table 6-1** Minor loss through meters

| Meter | Minimum Flow Rate (*gpm*) | Low Normal Flow Rate (*gpm*) | Changeover Range (Compound Meters) | High Normal Flow Rate (*gpm*) | Maximum Flow Rate (*gpm*) | Head Loss at Maximum Flow (*psi*) |
|---|---|---|---|---|---|---|
| Positive Displacement | | | | | | |
| ½ in. | 0.25 | 1 | N/A | 7.5 | 15 | 15 |
| ⅝ in. | 0.25 | 1 | | 10 | 20 | 15 |
| ¾ in. | 0.50 | 2 | | 15 | 30 | 15 |
| 1 in. | 0.75 | 3 | | 25 | 50 | 15 |
| 1½ in. | 1.50 | 5 | | 50 | 100 | 15 |
| 2 in. | 2.000 | 8 | | 80 | 160 | 15 |
| Multijet | | | | | | |
| ⅝ in. | 0.25 | 1 | N/A | 10 | 20 | 15 |
| ¾ in. | 0.500 | 2 | | 15 | 30 | 15 |
| 1 in. | 0.75 | 3 | | 25 | 50 | 15 |
| 1½ in. | 1.500 | 5 | | 50 | 100 | 15 |
| 2 in. | 2.000 | 8 | | 80 | 160 | 15 |
| Turbine class II | | | | | | |
| 1½ in. | N/A | 4 | N/A | 90 | 120 | 7 |
| 2 in. | | 4.00 | | 160 | 190 | 7 |
| 3 in. | | 8.00 | | 350 | 435 | 7 |
| 4 in. | | 15.00 | | 650 | 750 | 7 |
| 6 in. | | 30.00 | | 1,400 | 1,600 | 7 |
| 8 in. | | 50.00 | | 2,400 | 2,800 | 7 |
| 10 in. | | 75 | | 3,500 | 4,200 | 7 |
| 12 in. | | 120.00 | | 4,400 | 5,300 | 7 |
| 16 in. | | 200 | | 6,500 | 7,800 | 7 |
| 20 in. | | 300.00 | | 10,000 | 12,000 | 7 |
| Compound class II | | | | | | |
| 2 in. | 0.25 | 1 | 13 | 80 | 160 | 15 |
| 3 in. | 0.50 | 2 | 15 | 175 | 350 | 15 |
| 4 in. | 0.75 | 3 | 18 | 300 | 600 | 15 |
| 6 in. | 1.500 | 5 | 20 | 675 | 1,350 | 15 |
| 8 in. | 2.00 | 16 | 35 | 900 | 1,600 | 15 |
| Electromagnetic, Ultrasonic meters | | | | | | |
| Type I meters | | | | | | |
| ½ in. | 0.11 | 0.18 | N/A | 15 | 15 | 10 |
| ⅝ in. | 0.13 | 0.2 | | 20 | 20 | 10 |
| ¾ in. | 0.15 | 0.5 | | 30 | 30 | 10 |
| 1 in. | 0.30 | 0.75 | | 50 | 50 | 10 |
| 1½ in. | 0.60 | 2 | | 100 | 100 | 10 |
| 2 in. | 1.00 | 2.5 | | 160 | 160 | 10 |
| 3 in. | 2.50 | 7.5 | | 3,510 | 350 | 8 |

*(continued)*

**Table 6-1** Minor loss through meters (*Continued*)

| Meter | Minimum Flow Rate (*gpm*) | Low Normal Flow Rate (*gpm*) | Changeover Range (Compound Meters) | High Normal Flow Rate (*gpm*) | Maximum Flow Rate (*gpm*) | Head Loss at Maximum Flow (*psi*) |
|---|---|---|---|---|---|---|
| 4 in. | 3.50 | 10 | | 600 | 600 | 8 |
| 6 in. | 9.00 | 20 | | 1,350 | 1,350 | 8 |
| 8 in. | 18.00 | 40 | | 1,600 | 1,600 | 8 |
| Type II meters | | | | | | |
| ½ in. | 0.25 | 1 | N/A | 15 | 15 | 10 |
| ⅝ in. | 0.25 | 1 | | 20 | 20 | 10 |
| ¾ in. | 0.50 | 2 | | 30 | 30 | 10 |
| 1 in. | 0.75 | 3 | | 50 | 50 | 10 |
| 1½ in. | 1.50 | 4 | | 150 | 150 | 7 |
| 2 in. | 2.00 | 6 | | 220 | 220 | 7 |
| 3 in. | 4.00 | 15 | | 500 | 500 | 7 |
| 4 in. | 7.50 | 25 | | 880 | 880 | 7 |
| 6 in. | 15 | 55 | | 1,400 | 1,400 | 7 |
| 8 in. | 30 | 100 | | 2,800 | 2,800 | 7 |
| 10 in. | 50 | 150 | | 4,500 | 4,500 | 7 |
| 12 in. | 65 | 210 | | 5,500 | 5,500 | 7 |
| 16 in. | 110 | 375 | | 8,000 | 8,000 | 7 |
| 20 in. | 175 | 575 | | 12,000 | 12,000 | 7 |
| Fire service, type II—compound | | | | | | |
| 3 in. | * see note | 2 | 30 | 250 | 350 | 12 |
| 4 in. | | 4 | 40 | 400 | 700 | 12 |
| 6 in. | | 5 | 90 | 900 | 1,600 | 12 |
| 8 in. | | 8 | 150 | 1,600 | 2,800 | 12 |
| 10 in. | | 8 | 200 | 2,200 | 4,400 | 12 |
| Fire service, type III—turbine | | | | | | |
| 3 in. | 4 | 5 | N/A | 250 | 350 | 11 |
| 4 in. | 10 | 15 | | 400 | 700 | 11 |
| 6 in. | 20 | 30 | | 900 | 1,600 | 11 |
| 8 in. | 30 | 35 | | 1,600 | 2,800 | 11 |
| 10 in. | 35 | 55 | | 2,500 | 4,400 | 11 |
| Propeller (main line) | | | | | | |
| 2 in. | N/A | 45 | N/A | 100 | 120 | 5 |
| 3 in. | | 80 | | 250 | 300 | 5 |
| 4 in. | | 85 | | 500 | 600 | 2 |
| 6 in. | | 160 | | 1,200 | 1,350 | 1 |
| 8 in. | | 190 | | 1,500 | 1,800 | 0.5 |
| 10 in. | | 260 | | 2,000 | 2,400 | 0.5 |
| 12 in | | 275 | | 2,800 | 3,375 | 0.5 |
| 14 in. | | 350 | | 3,750 | 4,500 | 0.5 |
| 16 in. | | 450 | | 4,750 | 5,700 | 0.5 |

*(continued)*

**Table 6-1** Minor loss through meters (*Continued*)

| Meter | Minimum Flow Rate (*gpm*) | Low Normal Flow Rate (*gpm*) | Changeover Range (Compound Meters) | High Normal Flow Rate (*gpm*) | Maximum Flow Rate (*gpm*) | Head Loss at Maximum Flow (*psi*) |
|---|---|---|---|---|---|---|
| 18 in. | | 550 | | 5,625 | 6,750 | 0.25 |
| 20 in. | | 650 | | 6,875 | 8,250 | 0.25 |
| 24 in. | | 1,000 | | 10,000 | 12,000 | 0.25 |
| 30 in. | | 1,600 | | 15,000 | 18,000 | 0.25 |
| 36 in. | | 2,400 | | 20,000 | 24,000 | 0.25 |
| 42 in. | | 2,800 | | 28,000 | 40,000 | 0.1 |
| 48 in. | | 3,500 | | 35,000 | 50,000 | 0.1 |
| 54 in. | | 5,000 | | 45,000 | 55,000 | 0.1 |
| 60 in. | | 6,000 | | 60,000 | 80,000 | 0.1 |
| 66 in. | | 7,500 | | 75,000 | 95,000 | 0.1 |
| 72 in. | | 9,000 | | 90,000 | 115,000 | 0.1 |
| Fluidic Oscillator | | | | | | |
| ½ in. | 0.25 | 1 | N/A | 7.5 | 15 | 15 |
| ⅝ in. | 0.25 | 1 | | 10 | 20 | 15 |
| ¾ in. | 0.50 | 2 | | 15 | 30 | 15 |
| 1 in. | 0.75 | 3 | | 25 | 50 | 15 |
| 1½ in. | 1.50 | 5 | | 50 | 100 | 15 |
| 2 in. | 2.00 | 8 | | 80 | 160 | 15 |
| Singlejet | | | | | | |
| ⅝ in. | 0.25 | 1 | N/A | 10 | 20 | 15 |
| ¾ in. | 0.50 | 2 | | 15 | 30 | 15 |
| 1 in. | 0.75 | 3 | | 20 | 40 | 15 |
| 1½ in. | 0.50 | 1.5 | N/A | 50 | 100 | 15 |
| 2 in. | 0.50 | 2.0 | | 80 | 160 | 15 |
| 3 in. | 0.50 | 2.5 | | 160 | 320 | 15 |
| 4 in. | 0.75 | 3.0 | | 250 | 500 | 15 |
| 6 in | 1.50 | 4.0 | | 500 | 1,000 | 15 |
| Resid. Fire Sprinkler | | | | | | |
| ¾ in. | 0.5 | 2 | N/A | 15 | 30 | 10.1 |
| 1 in. | 0.75 | 2 | | 25 | 50 | 10.7 |
| 1½ in. | 1.5 | 3 | | 50 | 100 | 7.7 |
| 2 in. | 2.0 | 4 | | 80 | 160 | 7.7 |
| Resid. Fire Sprinkler w/ Strainer | | | | | | |
| ¾ in. | 0.5 | 2 | N/A | 15 | 30 | 14.5 |
| 1 in. | 0.75 | 2 | | 25 | 50 | 15.3 |
| 1 ½ in. | 1.5 | 3 | | 50 | 100 | 11 |
| 2 in. | 2.0 | 4 | | 80 | 160 | 11 |

*Source: Data are drawn from ANSI/AWWA Standards C700, C701, C702, C703, C704, C708, C710, C712, C713, C714, and C715 latest revision.*

N/A = not applicable

*Minimum flow rate is per the applicable AWWA standard for the bypass meter employed.

**Table 6-2** Typical uses for each type of meter classification

| Meter Type | Suggested Use |
|---|---|
| Positive displacement, Multijet, Fluidic oscillator, Singlejet, Electromagnetic, or Ultrasonic | Single-family residential, apartment buildings with fewer than 100 units |
| Positive displacement, Multijet, Singlejet, Turbine, Electromagnetic, or Ultrasonic | Small businesses (e.g., filling stations, restaurants, small hotels and motels, small office buildings, retail stores), schools, and other public buildings without large irrigation demands |
| Turbine, Singlejet, Electromagnetic, or Ultrasonic | Large hotels, factories, hospitals, irrigation, large office building, pump discharge, laundries, nursing homes |
| Compound, Singlejet, Turbine, Multijet, Electromagnetic, or Ultrasonic | Schools (with irrigation), apartment buildings with more than 100 units, dormitories, assisted living centers, retail shopping centers |
| Residential fire meters, typically 2-in. and smaller: Positive displacement, Electromagnetic, Ultrasonic, others | One and two-family dwellings and manufactured homes (NFPA 13D applications) |
| Fire line meters, typically 3-in. and larger | Fire service (for various NFPA 13 and NFPA 13R applications) |
| Differential pressure (venturi, orifice, flow tube), Turbine, Electromagnetic, or Ultrasonic | Pump discharge, wholesale water purchasers, research applications,subsystem metering |

maximum flow for this size meter. Weller et al. (2018) evaluated how various types of ⅝-in. × ¾-in. residential water meters respond to flow rates above the maximum rated flow rate (20 gpm) of these meters. Accuracy and pressure loss tests were conducted on a gravimetric test bench at flow rates of up to 2½ times the maximum rated flow rate. The results of this study show that the ultrasonic meters decreased significantly in accuracy for tests above 35 gpm. Nutating disk (ND) and oscillating piston (OP) meters were found to be accurate through nearly all surge flow rates tested, up to at least 65 gpm. Electromagnetic meters were found to be accurate up to 55 gpm. Comparing accuracy tests performed before and after surge flows showed that some ND and OP meters decreased in accuracy at flow rates below 2 gpm. Electromagnetic and ultrasonic meters had similar accuracy before and after surge flows. The Fire Protection Research Foundation (2011) conducted a separate research project, "Residential Fire Sprinklers—Water Usage and Water Meter Performance Study." This study also found that several types of residential water meters can sustain flows above their maximum rated flow successfully and without damage to the meter, particularly since the high flow rates are experienced infrequently and for short duration.

The designer should know the distribution of flow rates when sizing the meter for a customer. For example, if 99.0 percent of a customer's flow is below 30 gpm, but the customer has a brief spike (surge flow) once a week of 40 gpm, the meter should be sized to accurately collect the 99.0 percent of total volume through the meter, provided the head loss during the spike is acceptable and the meter can survive this surge flow. Many commonly used meters can sustain brief surge flows and still accurately record low flow rates, so it is not necessary to ignore 99.0 percent of the volume simply to keep the instantaneous peak demand within the meter's specified flow range. The distribution of flow rates is also important in determining what type of meter is preferred for larger services.

Although the selection of the type and size of meter should be based primarily on the range of flow and peak demand, pressure loss through the meter should also be a consideration. Newer turbine (e.g., "floating ball" technology), electromagnetic, and ultrasonic metering technologies are being extended into revenue billing applications at the retail level. These meters provide high accuracy at both high and low flows and offer more flexibility in addressing widely varying flow profiles.

If a meter is being sized for the first phase of a larger project, the meter could be sized for the initial phase with provision for the installation of a larger meter in the future when expansion occurs. Similarly, if the designer anticipates a greater demand by the customer at some point in the future, provision should be made to enable the installation of a larger meter. Again, the meter could be installed for current needs but with a meter box and connections that are adequate to meet anticipated future requirements.

Positive displacement meters are widely used because they accurately register at the M6 low flow test level (AWWA 2012). Static (ultrasonic and electromagnetic) meters are capable of accurately measuring at the C715 ultra-low flow test level. Multijet, singlejet, fluidic oscillator, electromagnetic, and ultrasonic meters have been used in some applications similar to those that employ positive displacement meters. Traditionally, turbine and propeller meters have been used when water usage is characterized by higher demands with few or no low flows because these meters are not designed to register low flows as accurately as other meter types. Compound meters are a combination of a smaller bypass meter and a larger (main-line) turbine-type meter. They register accurately across the largest range of flow rates. However, compound meters have higher wear and maintenance requirements and should be carefully selected to provide economical service. Compound water meters have two separate registers, which require separate reading and combination for customer billing. Both registers need to be accounted for and attributed to a single meter and single customer. Also, compound meters may be less accurate if flows occur frequently at the "cross-over" flow range, or the range where the flows begin to be diverted to the turbine meter side. To maintain accurate readings, meters should be tested regularly (AWWA 2012, Chapters 5 and 6) and replaced as necessary.

Critical customers, such as hospitals, apartment complexes, nursing homes, certain schools, and commercial and industrial applications with only a single water service connection line and processes requiring uninterrupted water service, should have bypasses installed around the meter so that meter test and repair activities can be performed at scheduled intervals without inconvenience to either the customer or the utility. The bypass line should be valved and locked appropriately. If the level of service is not critical, it is permissible to make the bypass line one size smaller than the service line as a cost-savings measure. If unauthorized consumption is a concern, the bypass line should be metered.

Additionally, plate strainers should be installed in front of turbine, large singlejet, and compound meters. In some cases, these strainers are an integral part of the meter housing. The designer should account for the head loss from the strainer in the calculations. The selection of the type of meter is discussed further in M6 (AWWA 2012). Typical uses are shown in Table 6-2 for the various types of meters.

## METER ARRANGEMENTS FOR FIRE SERVICES

One of the key decisions in sizing service lines and meters is the arrangement of service lines and meters. Some utilities use a single tapped line to supply both fire and domestic use (including commercial/industrial), but other utilities use a separately tapped line for domestic service and a separately tapped line for fire use. Also, for fire lines, some utilities provide a water meter on the line; some use a detector check-device to detect water flow; and others include no equipment. The most common arrangements for service lines and meters are shown in Figures 6-1 through 6-4.

The decision as to which service line and meter arrangement to use is a policy decision of the utility, with recognition of the provisions of local fire, plumbing, and building codes.

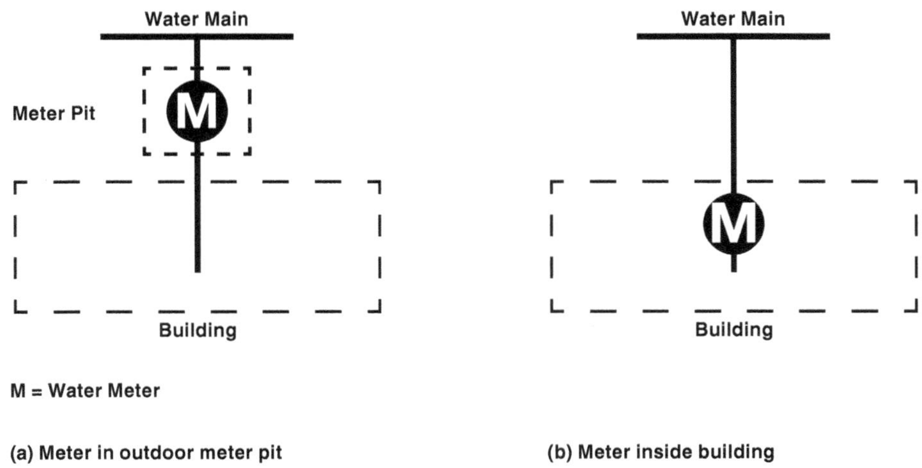


**Figure 6-1** Building served by domestic water supply only; no fire connection exists

Metering fire flow (either with a meter on the fire line or a meter on the common line supplying domestic and fire service) maximizes water accountability. Use of a common line meter requires a multiregister meter in some commercial/industrial applications, but a separate fire line makes it possible to function with a single fire-type turbine meter on the fire line in these same applications. In some utilities, a policy exists to provide fire flow at no cost (or a flat fee), and fire flow need not be metered. In those cases, the utility may still want to use a detector check to minimize the likelihood of unauthorized consumption from the fire line and to identify the existence of leakage in the fire line.

Figure 6-1 shows two configurations used to supply water service to a building without fire sprinklers. In these cases, the meter can be installed in an outdoor meter pit (left graphic) or inside the building (right graphic). These configurations are the most common ones in use in the water industry since most buildings do not have fire lines.

Figures 6-2 through 6-4 show the arrangement options for buildings that are provided fire supply. Figure 6-2 shows six arrangements that are typically used in fire sprinkler systems for larger buildings designed under the NFPA 13 (NFPA 2019a) and NFPA 13R (NFPA 2019b) standards. These systems, which feature a separately tapped domestic supply line and separate fire line, are most common in commercial, industrial, and multifamily housing buildings. The top two arrangements in Figure 6-2 show metered (M) domestic lines but unmetered fire lines—a common configuration. An unmetered fire configuration is often justified because of a perceived high cost of a meter when water use seldomly occurs in fire lines. However, unmetered fire lines can be at risk of unauthorized consumption if a user taps domestic piping into the fire line. As water is drawn through the fire line, metering does not exist to measure the water use and allow the utility to bill for it.

The middle arrangements in Figure 6-2 show a metered domestic line along with a fire line with a detector check (D) installed. The detector check detects flow in the line without measuring the flow rate. This device can alert the utility to water use through the line, thereby detecting unauthorized consumption or leakage, and it is less expensive than the cost of a fire meter.

The bottom two arrangements shown in Figure 6-2 show a water meter installed on both the domestic line and the fire line. Metering fire lines provides a measure of flow through the line when activated and is a means to detect unauthorized consumption. Fire-rated water meters, however, come at a cost. Many water meter manufacturers produce meters with strainers specifically designed for fire service. Fire insurance

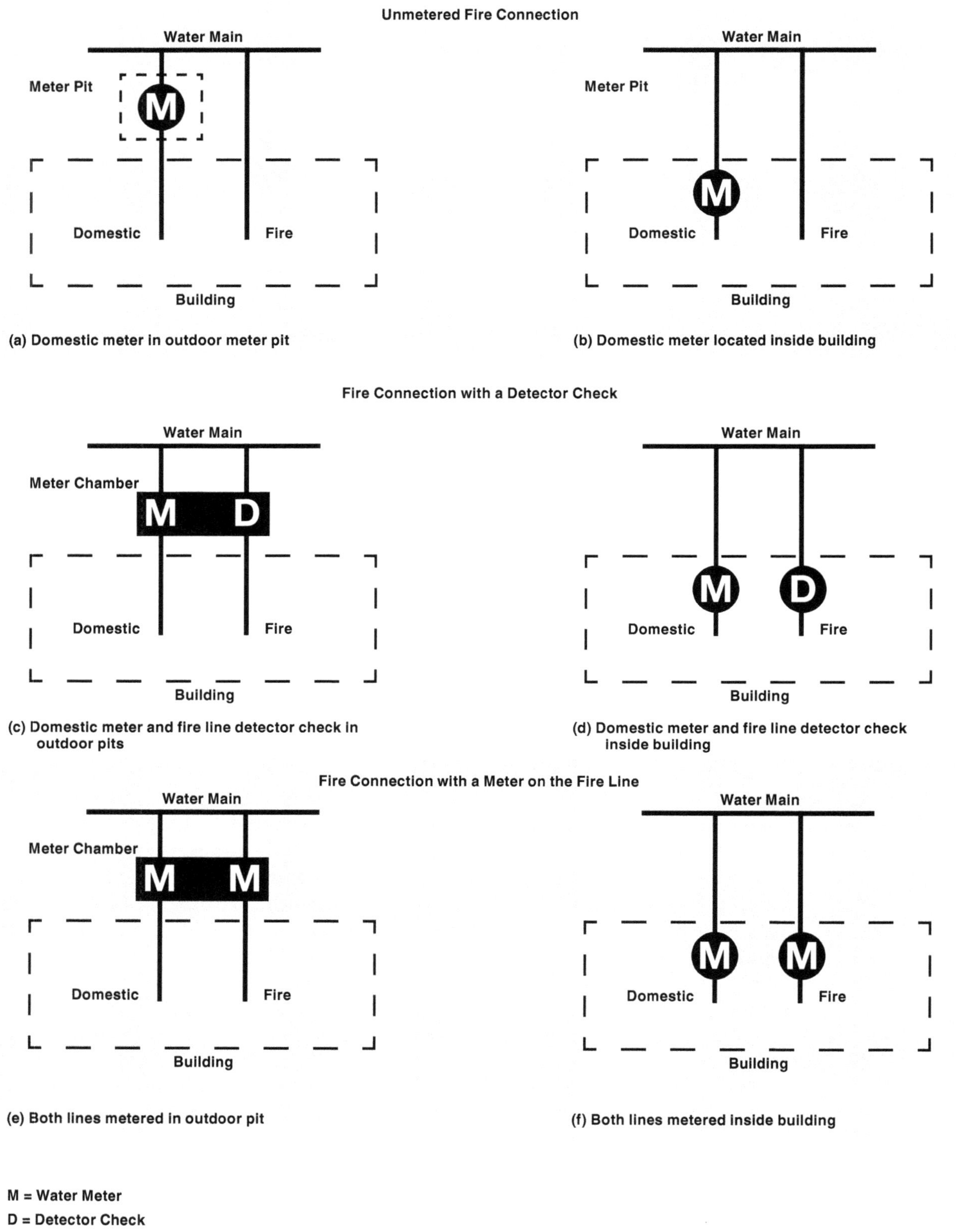


Figure 6-2 Building supplied by domestic water supply and a separate fire connection; typically commercial, industrial, or larger multifamily housing designed under the NFPA 13 or 13R standards

companies recognize turbine meters and turbine meters with a check-valve bypass meter configuration that are equipped with UL-Listed or FM-Approved strainers. While these meter designs are appropriate for water supplied to fire service, they represent a greater cost than equivalently sized meters designed for domestic water service.

The arrangements shown in Figures 6-3 and 6-4 give the options that exist for Residential Fire Sprinkler Systems (RFSS) typically designed under the 2019 NFPA 13D

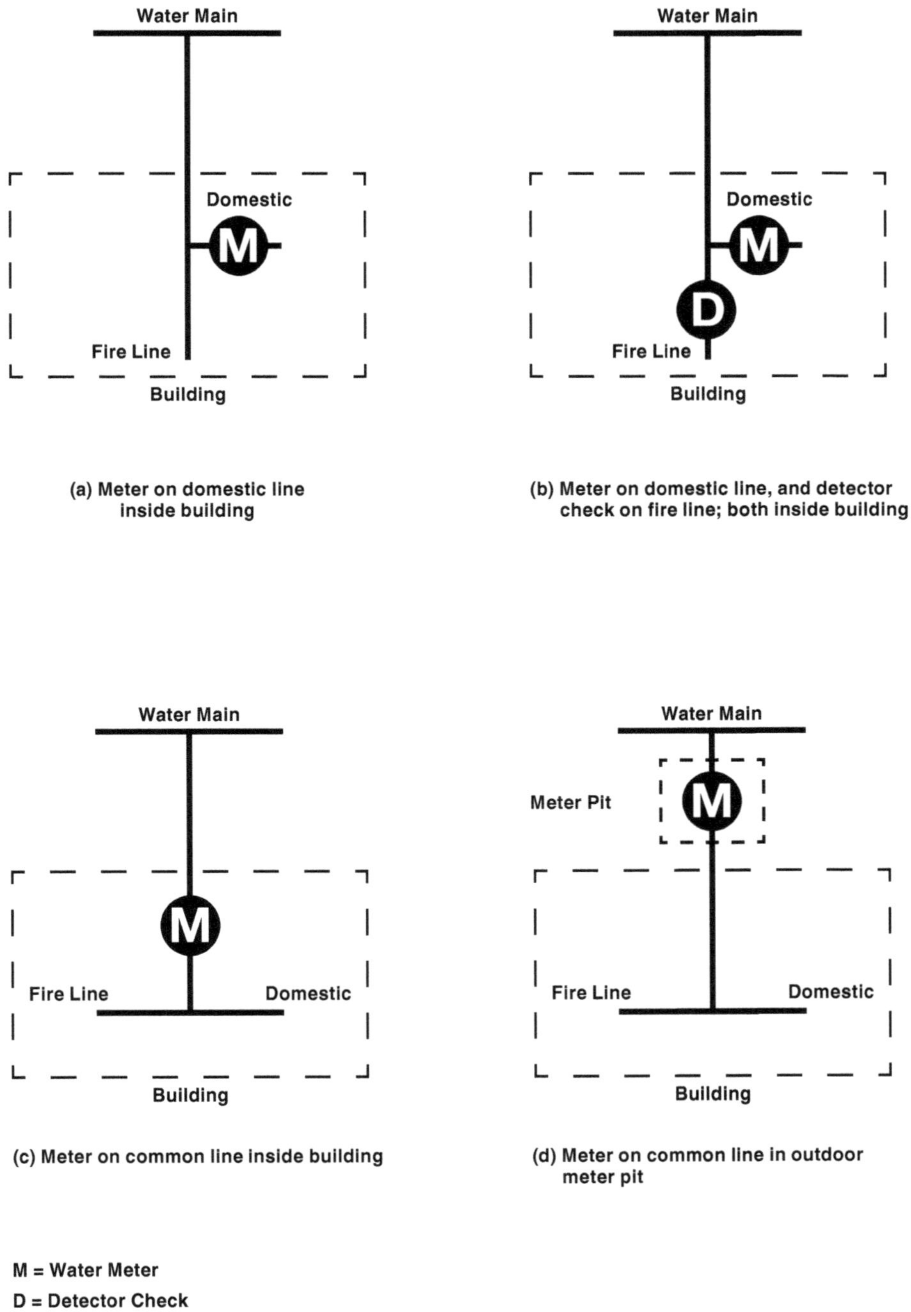


Figure 6-3 One- or two-story residential building outfitted with a Residential Fire Sprinkler System (RFSS) typically designed under the NFPA 13D Standard—Stand-alone configurations with a single line tapped into the water main

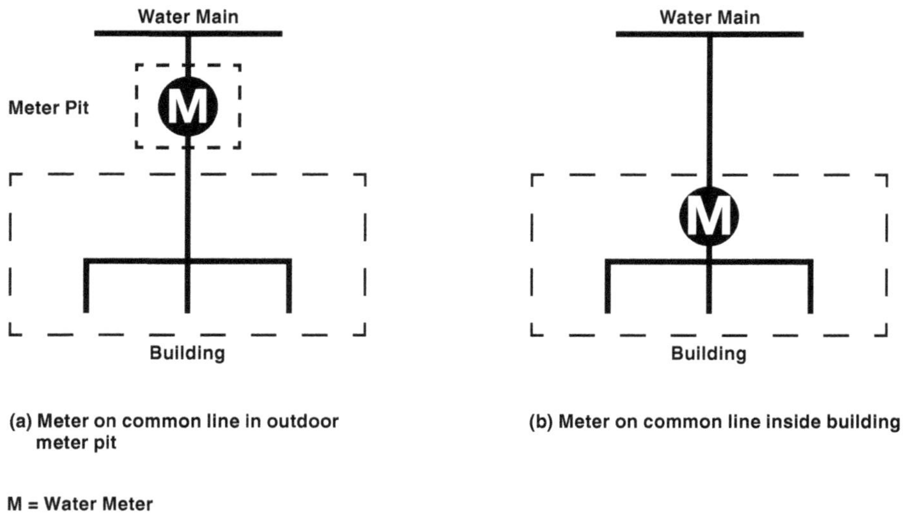


**Figure 6-4 One- or two-story residential building outfitted with a Residential Fire Sprinkler System (RFSS) typically designed under the NFPA 13D Standard—Multipurpose piping configuration. In this configuration, a single line is tapped into the water main, and this line enters the building and extends to provide water to both domestic water fixtures and fire sprinkler heads through common plumbing piping.**

Standard for the Installation of Sprinkler Systems in One- and Two-Family Dwellings and Manufactured Homes (NFPA 2019c). Meter sizes of ¾-in. or 1-in. are sufficient in many RFSS applications. The use of other meter sizes should be approached with caution but may be appropriate when considering the available pressure and expected water demands at a given location. For residential applications, Underwriters Laboratories has issued residential fire-service listings for various metering technologies from several manufacturers, although NFPA 13D does not require a UL-listed product.

Two options exist under the NFPA 13D Standard for piping configurations inside buildings—*Stand-alone* and *Multipurpose*—and these are explained as follows.

Figure 6-3 shows four stand-alone configurations under the NFPA 13D Standard, all of which emanate from a single tapped line from the water main. Note that the NFPA 13D Standard also allows for a separately tapped fire line with options similar to those shown in Figure 6-2; however, these options are not repeated in Figure 6-3, since separate fire lines in smaller residential occupancies represent a significant cost increase relative to the arrangements shown in Figure 6-3. In the stand-alone configurations in Figure 6-3, internal piping branches inside the building to supply domestic fixtures with one branch and fire sprinkler heads with the other branch. In the top two arrangements in Figure 6-3, only the domestic flow is metered, and the fire line is unmetered. In the top-right illustration, a detector check is installed on the fire line.

In the bottom two arrangements in Figure 6-3, the water meter is installed on the common piping upstream of the branch between the domestic and fire supplies. In this way, all flow is registered by the water meter, whether flowing through the domestic line or the fire line.

Figure 6-4 shows two arrangements possible in the multipurpose configuration. In this configuration, the same piping downstream of the water meter is routed to supply both domestic fixtures and fire sprinkler heads. See the 13D Standard (NFPA 2019c) and the publication *Residential Fire Sprinkler Systems: Guidance for Water Utilities* (AWWA 2018) for more information on RFSS applications.

## SUMMARY

The meter should be correctly sized to handle and record the expected range of flow rates. Utility administrative policies should encourage good engineering and design. Undersizing the meter can cause pressure-related problems and can also result in lost revenue at high flow rates. Oversizing the meter means overspending for the meter and can result in reduced revenue at low flow rates and inaccurate meter recordings. In all cases, the meter should be accurately sized to meet the customer's actual or projected demands and the utility's engineering criteria. AWWA meter standards and meter manufacturer specifications are the recommended guidelines for determining the correct meter type and size once anticipated or actual demand characteristics have been defined using either flow recorders or manual fixture value calculations.

## REFERENCES


American National Standards Institute/American Water Works Association (ANSI/AWWA). C700. Cold-Water Meters—Displacement Type, Bronze Main Case. Denver: AWWA.

ANSI/AWWA. C701. Cold-Water Meters—Turbine Type, for Customer Service. Denver: AWWA.

ANSI/AWWA. C702. Cold-Water Meters—Compound Type. Denver: AWWA.

ANSI/AWWA. C703. Cold-Water Meters—Fire-Service Type. Denver: AWWA.

ANSI/AWWA. C704. Propeller-Type Meters for Waterworks Applications. Denver: AWWA.

ANSI/AWWA. C706. Direct-Reading, Remote-Registration Systems for Cold-Water Meters. Denver: AWWA.

ANSI/AWWA. C707. Encoder-Type Remote-Registration Systems for Cold-Water Meters. Denver: AWWA.

ANSI/AWWA. C708. Cold-Water Meters—Multijet Type. Denver: AWWA.

ANSI/AWWA. C710. Cold-Water Meters—Displacement Type, Plastic Main Case. Denver: AWWA.

ANSI/AWWA. C712. Cold-Water Meters—Singlejet Type. Denver: AWWA.

ANSI/AWWA. C713. Cold-Water Meters—Fluidic-Oscillator Type. Denver: AWWA.

ANSI/AWWA. C714. Cold-Water Meters for Residential Fire Sprinkler Systems in One- and Two-Family Dwellings and Manufactured Homes. Denver: AWWA.

ANSI/AWWA. C715. Cold-Water Meters—Electromagnetic and Ultrasonic Type for Revenue Applications. Denver: AWWA.

AWWA. 2012 (5th ed.). M6, *Water Meters—Selection, Installation, Testing, and Maintenance*. Denver: AWWA.

AWWA. 2018. "Residential Fire Sprinkler Systems: Guidance for Water Utilities." Denver: AWWA.

Fire Protection Research Foundation. 2011. "Residential Fire Sprinklers—Water Usage and Water Meter Performance Study." Quincy, Mass.: National Fire Protection Association.

National Fire Protection Association (NFPA). 2019a. NFPA 13, Standard for the Installation of Sprinkler Systems. Quincy, Mass.: NFPA.

NFPA. 2019b. NFPA 13R, Standard for the Installation of Sprinkler Systems in Low-Rise Residential Occupancies. Quincy, Mass.: NFPA.

NFPA. 2019c. NFPA 13D, Standard for the Installation of Sprinkler Systems in One- and Two-Family Dwellings and Manufactured Homes. Quincy, Mass.: NFPA.

Sumrak, M.; Johnson, M.; and Barfuss, S. "Comparing Low-Flow Accuracy of Mechanical and Electronic Meters." *Journal AWWA*. 108(6):E327–E334.

Weller, R.; Barfuss, S.; and Johnson, M. 2018. "The Effect of Surge Flows on Residential Water Meters." *AWWA Water Science*. 1(1):e1117.

This page intentionally blank.

Appendix A

# Software to Size a Meter

## INTRODUCTION

The most accurate way to size a meter is to record detailed demand data for a particular customer or a similar customer during a period of typical and repeatable usage, and use that flow rate data, the meter manufacturer's specification sheets, and the water rate structure (and sewer rate structure where applicable) to determine the:

- accounted-for water volume,
- percentage of flow within a meter's normal range,
- head loss at the maximum recorded flow rate, and
- revenue impact for each possible meter size and type under consideration.

Because such calculations involve many variables and are too time-consuming to perform both correctly and efficiently by hand, the evaluator should take advantage of digital spreadsheets and/or specialized meter-sizing software to perform these calculations. This appendix illustrates how demand profile data and meter performance characteristics can be combined in meter-sizing software to aid in the selection of meter type and size for a service.

The following is an overview of how meter-sizing software can use demand data in conjunction with meter performance specifications and utility-specific meter and financial information to perform the calculations on which proper meter sizing and typing decisions are based. Once utility-specific information has been entered, the software can automatically estimate the projected volume of accounted-for water and the annual revenue increase (or decrease) associated with each meter type and size option. It can also calculate the percentage of flow within the normal flow range of each meter option; the number of months it would take for a meter replacement to pay for itself; and the head loss at the maximum recorded flow for each meter option (except for compound meters). The software accomplishes this by distributing high-resolution demand data collected at a meter site into software-defined flow ranges for each meter type and size option.

## SUMMARY OF SOFTWARE OPERATION

Before beginning an iteration, the user should confirm or change the software defaults and enter the meter and utility-specific information. Some software uses AWWA meter flow ranges and accuracy factors for positive displacement, multijet, Class I and Class II turbo, and compound meters as found in AWWA water meter standards. The software should enable the user to edit these values when appropriate. Once the correct values have been entered, the user imports a demand profile made using a flow recorder (i.e., a profile that provides flow rate through the meter versus time on some small interval such as 10 s). This data can be entered manually; however, much of the advantage of using the software comes from its capacity to base its calculations on high-resolution rate information gathered over a period of time. It is imperative that the operator enters the correct size and type of meter from which the data was collected.

If needed, a seasonal adjustment can be performed by entering the billed usage and the days in the billing cycle for each billing cycle during the previous year, and the flow rates will be scaled up or down proportionately. Use care when entering the billed usage. The utility may bill in thousands of gallons or hundreds of cubic feet, but the usage data will need to be entered in gallons or cubic feet. Seasonal adjustments are generally used at facilities that may be vacant for part of the year, such as for schools, resort properties, etc., that may be closed or have greatly reduced occupancy during part of the year. Seasonal adjustments are not needed whenever the usage pattern remains relatively consistent year-round.

At this point, sufficient data has been entered to perform calculations. The software can rank meters based on accounted-for water and determine the amount of flow in each meter's normal range; the head loss at the maximum recorded flow rate (except for compound water meters); and the financial impact of each meter alternative. A graphic analysis feature can enable a user to view each meter's maximum, minimum, high-normal, and low-normal flow rates superimposed over the demand profile for a given installation.

## SETUP

Current meter-sizing software is designed around AWWA C700-series meter performance standards and comes with defaults that permit a meter-sizing analysis to be performed without altering the default AWWA parameters (Figure A-1). Nevertheless, the defaults and software assumptions must be reviewed before making any sizing decisions based on software results. Certain defaults, such as the rates of flow applicable to compound meter change-over ranges, do not have AWWA standard values; accordingly, software should use reasonable default values based on consultation with industry professionals.

The software should enable all assumptions to be tailored to meet the unique requirements and characteristics of any utility. Many utilities have their own standards of meter performance, which represent local conditions or tests, and each utility has a unique financial structure. The minimum setup requirements for performing a financial analysis are:

- the retail cost of water (and related wastewater rates, if applicable);
- fixed charges for each meter type and size;
- changeout costs for each meter type and size (labor and materials); and

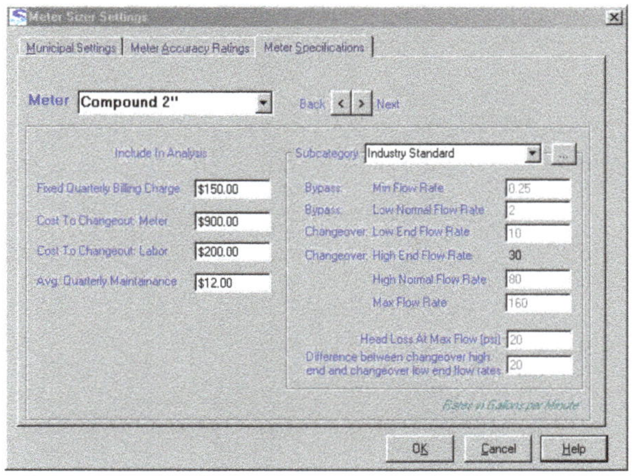


*Source:* Courtesy of F.S. Brainard and Co.

**Figure A-1 Meter specifications screen (sample from MeterSizer®)**

- the average monthly maintenance charge for each meter type and size. For example, compound meters typically have higher maintenance costs than turbines, and the analysis should reflect this.

A software program should include information about each meter type and size's accuracy in various flow ranges. AWWA standards can be used as defaults for these values, although information provided by manufacturers may be more appropriate. These values are:

- the maximum and minimum flow rates;
- the flow rates defining the normal flow range;
- the changeover flow range (for compound meters);
- the head loss at the maximum flow rate; and
- accounted-for meter accuracy ratings for each applicable flow range.

Because meters do not stop registering flow for rates above and below AWWA standard accuracy ranges, software should also make assumptions for performance in these ranges. Default values for these assumptions are developed in consultation with industry professionals. While a user may change these assumptions by adding other categories with their own values, the software default should be used unless the user has a good basis for using alternate values.

## SIZING A METER

Once all the meter performance and utility financial information has been entered/confirmed, sizing a meter is a seven-step procedure.

### 1. Site Survey

If possible, a questionnaire should be completed to ensure that the period chosen for recording the consumption profile is representative of the user's typical usage patterns. If it is not, the field test should be rescheduled for a more appropriate time period. For example, it is not advisable to install flow records at a school during the summer. Peak flow events should also be captured. If such events are seasonal, the customer should be profiled during normal demand and peak demand periods. For example, food processing

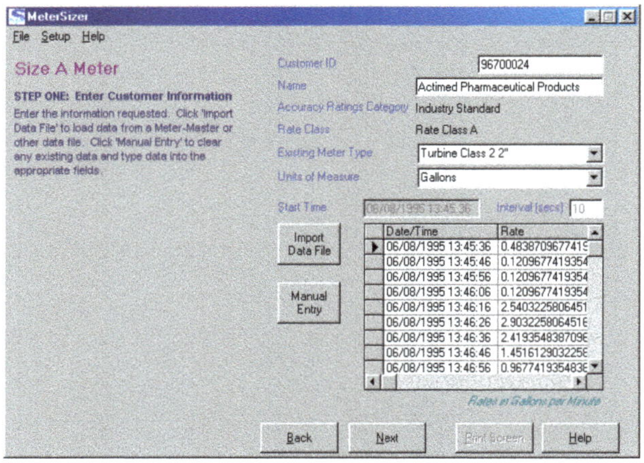


*Source:* Courtesy of F.S. Brainard and Co.

**Figure A-2 Customer information screen (sample from MeterSizer®)**

plants work all year but are busiest during harvest time. Usage data from a customer's billing history can also be used to verify that the amount of water used during the data-storage period is representative. A questionnaire can also be used to identify the potential sources of peak and minimum flows.

## 2. Consumption Profile

A consumption profile should be recorded using a data logger/recording device and then imported into the software. A minimum of 24 h is recommended for residential applications. Longer periods are recommended for commercial/industrial accounts because significant uses of water may only occur on certain days of the week.

## 3. Import/Enter Customer Information

Customer information and the actual demand data are imported into the software or entered directly by typing into screen fields (Figure A-2).

## 4. Seasonal Adjustment

To adjust data for seasonal variations in water use, an option can be provided on another screen to enter the customer's most recent annual billing history. This data enables software to consider the impact of changes in seasonal consumption patterns in the meter-sizing decision if it is reasonable to assume that the pattern in different seasons is consistent, with only the quantity of water used varying. If the average daily usage during the maximum-usage billing period is 20 percent higher than during the data-recording period, all rate values could be increased 20 percent, for example.

## 5. Initial Results

As an example of displaying results, *MeterSizer*® automatically shows the accounted-for and financial impacts associated with the best meter option for the application (Figure A-3). Left/Right arrows and a scroll bar allow the user to view results for all meter options. Meters are ranked from best-suited to worst-suited for the application based upon the volume of accounted-for water. If there are two meters with the same percentage, the program will pick the meter with the greatest percentage of flow in the normal range.

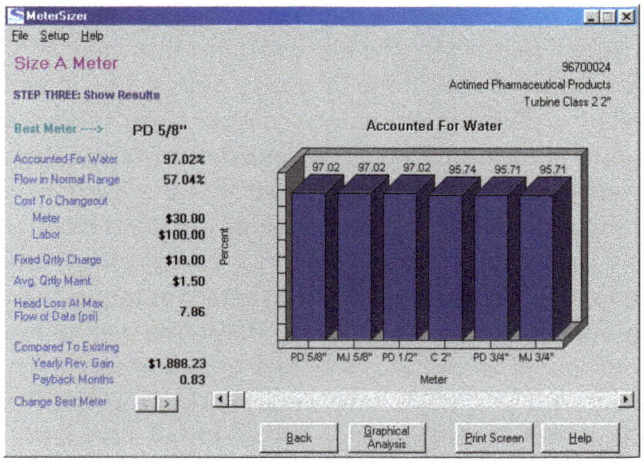


*Source:* Courtesy of F.S. Brainard and Co.

**Figure A-3 Initial results screen (sample from MeterSizer®)**

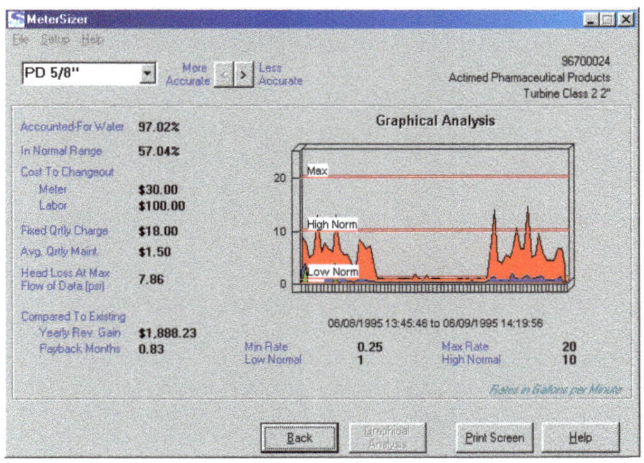


*Source:* Courtesy of F.S. Brainard and Co.

**Figure A-4 Graphical analysis screen (sample from MeterSizer®)**

If this, too, is a tie, it will default to the meter with the lowest head loss at maximum-recorded flow. Each meter option's projected performance can be further analyzed on the graphical analysis screen.

## 6. Graphical Analysis

The actual demand data can be viewed graphically against the specified flow ranges for each meter type and size option (Figure A-4). This enables the user to determine whether a recommended meter option is too small based on peak flow and the percentage of flow outside a meter's normal (optimal) flow range.

## 7. Calibration of Results

Meter-sizing software permits complex and time-consuming calculations to be performed quickly and accurately. It is only a tool, however, to assist the user in a final decision that often must take into consideration other factors, such as fire-flow requirements and possible changes in demand due to many possible causes. Also, no model is perfect. If the users are unsure about any of the assumptions, they may run the analysis using best- and worst-case scenarios to determine the range of possible outcomes. For example, if they are unsure of relative meter maintenance costs or wish to vary assumptions for accounted-for water below minimum and above maximum flow rates, they may run the analysis using different values. It is recommended that the users try different scenarios to better understand the potential deviation of results obtained through any model.

In many cases, the potential benefit for downsizing is overlooked, and designers will opt for a meter smaller than the existing meter yet larger than the optimum calculated by a software model. Meter sizing software allows the user to evaluate all potential scenarios and their consequences. In cases where a more conservative decision is either preferred or required because of specific considerations such as fire flows, the program provides a good indication of the cost (lost-revenue potential) associated with such a decision.

# Appendix B

# Flow Friction Loss Tables

The following tables (Tables B-1 through B-39) contain friction loss values for common service line materials (copper, polyethylene, PEX, and PVC) and sizes (0.75–8-in. diameter). The friction loss values are given in pounds per square inch pressure loss per 100 ft of pipe. The tables include only standard inside diameters. For other pipe materials and sizes, the friction loss values can be calculated from the Hazen-Williams formula by substituting the corresponding constants and values.

The tables contain the velocity of water in feet per second for corresponding head loss values that will be required in selecting the proper size service for the flow. High velocities can cause excessive water hammer (when valves are closed rapidly) and, in some cases, cavitation damage to service material. It is recommended that the maximum water velocity in the service line be not more than 10 ft/s. Higher values are listed in the tables for the engineer's use; however, these do not constitute a recommendation for the use of high velocities.

**Table B-1** Copper water tube properties—C=130—¾ in. nominal diameter

| *C* = 130 | ID = Inside diameter* | | | | | |
|---|---|---|---|---|---|---|
| | Type K<br>ID = 0.745 | | Type L<br>ID = 0.785 | | Type M<br>ID = 0.811 | |
| Flow<br>*gpm* | Head Loss<br>*ft/100 ft* | Velocity<br>*ft/s* | Head Loss<br>*ft/100 ft* | Velocity<br>*ft/s* | Head Loss<br>*ft/100 ft* | Velocity<br>*ft/s* |
| 1 | 0.53 | 0.74 | 0.41 | 0.66 | 0.35 | 0.62 |
| 2 | 1.92 | 1.47 | 1.49 | 1.33 | 1.27 | 1.24 |
| 3 | 4.07 | 2.21 | 3.16 | 1.99 | 2.69 | 1.86 |
| 4 | 6.93 | 2.94 | 5.38 | 2.65 | 4.59 | 2.48 |
| 5 | 10.48 | 3.68 | 8.13 | 3.31 | 6.93 | 3.11 |
| 6 | 14.69 | 4.42 | 11.39 | 3.98 | 9.72 | 3.73 |
| 7 | 19.55 | 5.15 | 15.15 | 4.64 | 12.93 | 4.35 |
| 8 | 25.03 | 5.89 | 19.40 | 5.30 | 16.55 | 4.97 |
| 9 | 31.13 | 6.62 | 24.13 | 5.97 | 20.59 | 5.59 |
| 10 | 37.84 | 7.36 | 29.33 | 6.63 | 25.03 | 6.21 |
| 11 | 45.14 | 8.10 | 34.99 | 7.29 | 29.86 | 6.83 |
| 12 | 53.04 | 8.83 | 41.11 | 7.95 | 35.08 | 7.45 |
| 13 | 61.51 | 9.57 | 47.68 | 8.62 | 40.68 | 8.07 |
| 14 | 70.56 | 10.30 | 54.69 | 9.28 | 46.67 | 8.70 |
| 15 | | | 62.14 | 9.94 | 53.03 | 9.32 |
| 16 | | | 70.03 | 10.61 | 59.76 | 9.94 |
| 17 | | | | | 66.86 | 10.56 |

C—roughness coefficient
*ASTM B88

**Table B-2** Copper water tube properties—C=130—1 in. nominal diameter

| *C* = 130 | ID = Inside diameter* | | | | | |
|---|---|---|---|---|---|---|
| | Type K<br>ID = 0.995 | | Type L<br>ID = 1.025 | | Type M<br>ID = 1.055 | |
| Flow<br>*gpm* | Head Loss<br>*ft/100 ft* | Velocity<br>*ft/s* | Head Loss<br>*ft/100 ft* | Velocity<br>*ft/s* | Head Loss<br>*ft/100 ft* | Velocity<br>*ft/s* |
| 2 | 0.47 | 0.83 | 0.41 | 0.78 | 0.35 | 0.73 |
| 3 | 0.99 | 1.24 | 0.86 | 1.17 | 0.75 | 1.10 |
| 4 | 1.69 | 1.65 | 1.47 | 1.56 | 1.27 | 1.47 |
| 5 | 2.56 | 2.06 | 2.22 | 1.94 | 1.93 | 1.84 |
| 6 | 3.59 | 2.48 | 3.11 | 2.33 | 2.70 | 2.20 |
| 7 | 4.78 | 2.89 | 4.13 | 2.72 | 3.59 | 2.57 |
| 8 | 6.12 | 3.30 | 5.29 | 3.11 | 4.60 | 2.94 |
| 9 | 7.61 | 3.71 | 6.58 | 3.50 | 5.72 | 3.30 |
| 10 | 9.24 | 4.13 | 8.00 | 3.89 | 6.95 | 3.67 |
| 12 | 12.96 | 4.95 | 11.21 | 4.67 | 9.74 | 4.40 |
| 14 | 17.24 | 5.78 | 14.92 | 5.44 | 12.96 | 5.14 |
| 16 | 22.07 | 6.60 | 19.10 | 6.22 | 16.60 | 5.87 |
| 18 | 27.45 | 7.43 | 23.76 | 7.00 | 20.64 | 6.61 |
| 20 | 33.37 | 8.25 | 28.87 | 7.78 | 25.09 | 7.34 |
| 25 | 50.44 | 10.32 | 43.65 | 9.72 | 37.93 | 9.18 |
| 30 | | | 61.18 | 11.66 | 53.16 | 11.01 |

*ASTM B88

**Table B-3** Copper water tube properties—C=130—1¼ in. nominal diameter

| *C* = 130 | ID = Inside diameter* | | | | | |
|---|---|---|---|---|---|---|
| | Type K<br>ID = 1.245 | | Type L<br>ID = 1.265 | | Type M<br>ID = 1.291 | |
| Flow<br>*gpm* | Head Loss<br>*ft/100 ft* | Velocity<br>*ft/s* | Head Loss<br>*ft/100 ft* | Velocity<br>*ft/s* | Head Loss<br>*ft/100 ft* | Velocity<br>*ft/s* |
| 5 | 0.86 | 1.32 | 0.80 | 1.28 | 0.72 | 1.23 |
| 6 | 1.20 | 1.58 | 1.11 | 1.53 | 1.01 | 1.47 |
| 7 | 1.60 | 1.84 | 1.48 | 1.79 | 1.34 | 1.72 |
| 8 | 2.05 | 2.11 | 1.90 | 2.04 | 1.72 | 1.96 |
| 9 | 2.55 | 2.37 | 2.36 | 2.30 | 2.14 | 2.21 |
| 10 | 3.10 | 2.64 | 2.87 | 2.55 | 2.60 | 2.45 |
| 12 | 4.35 | 3.16 | 4.02 | 3.06 | 3.64 | 2.94 |
| 15 | 6.57 | 3.95 | 6.08 | 3.83 | 5.51 | 3.68 |
| 20 | 11.20 | 5.27 | 10.36 | 5.11 | 9.39 | 4.90 |
| 25 | 16.93 | 6.59 | 15.67 | 6.38 | 14.19 | 6.13 |
| 30 | 23.73 | 7.91 | 21.96 | 7.66 | 19.89 | 7.35 |
| 35 | 31.57 | 9.22 | 29.21 | 8.93 | 26.46 | 8.58 |
| 40 | 40.43 | 10.54 | 37.41 | 10.21 | 33.88 | 9.80 |
| 45 | | | | | 42.14 | 11.03 |

*ASTM B88

**Table B-4** Copper water tube properties—C=130—1½ in. nominal diameter

| *C* = 130 | ID = Inside diameter* | | | | | |
|---|---|---|---|---|---|---|
| | Type K<br>ID = 1.481 | | Type L<br>ID = 1.505 | | Type M<br>ID = 1.527 | |
| Flow<br>*gpm* | Head Loss<br>*ft/100 ft* | Velocity<br>*ft/s* | Head Loss<br>*ft/100 ft* | Velocity<br>*ft/s* | Head Loss<br>*ft/100 ft* | Velocity<br>*ft/s* |
| 8 | 0.88 | 1.49 | 0.81 | 1.44 | 0.76 | 1.40 |
| 9 | 1.10 | 1.68 | 1.01 | 1.62 | 0.94 | 1.58 |
| 10 | 1.33 | 1.86 | 1.23 | 1.80 | 1.15 | 1.75 |
| 12 | 1.87 | 2.23 | 1.73 | 2.16 | 1.61 | 2.10 |
| 15 | 2.82 | 2.79 | 2.61 | 2.71 | 2.43 | 2.63 |
| 20 | 4.81 | 3.72 | 4.45 | 3.61 | 4.14 | 3.50 |
| 25 | 7.27 | 4.66 | 6.72 | 4.51 | 6.26 | 4.38 |
| 30 | 10.19 | 5.59 | 9.42 | 5.41 | 8.78 | 5.26 |
| 35 | 13.56 | 6.52 | 12.54 | 6.31 | 11.68 | 6.13 |
| 40 | 17.36 | 7.45 | 16.05 | 7.21 | 14.96 | 7.01 |
| 45 | 21.59 | 8.38 | 19.97 | 8.12 | 18.60 | 7.88 |
| 50 | 26.24 | 9.31 | 24.27 | 9.02 | 22.61 | 8.76 |
| 60 | 36.78 | 11.17 | 34.01 | 10.82 | 31.69 | 10.51 |

*ASTM B88

**Table B-5** Copper water tube properties—C=130—2 in. nominal diameter

| *C* = 130 | ID = Inside diameter* | | | | | |
|---|---|---|---|---|---|---|
| | Type K<br>ID = 1.959 | | Type L<br>ID = 1.985 | | Type M<br>ID = 2.009 | |
| Flow<br>*gpm* | Head Loss<br>*ft/100 ft* | Velocity<br>*ft/s* | Head Loss<br>*ft/100 ft* | Velocity<br>*ft/s* | Head Loss<br>*ft/100 ft* | Velocity<br>*ft/s* |
| 10 | 0.34 | 1.06 | 0.32 | 1.04 | 0.30 | 1.01 |
| 12 | 0.48 | 1.28 | 0.45 | 1.24 | 0.42 | 1.21 |
| 14 | 0.64 | 1.49 | 0.60 | 1.45 | 0.56 | 1.42 |
| 16 | 0.81 | 1.70 | 0.76 | 1.66 | 0.72 | 1.62 |
| 18 | 1.01 | 1.92 | 0.95 | 1.87 | 0.90 | 1.82 |
| 20 | 1.23 | 2.13 | 1.15 | 2.07 | 1.09 | 2.02 |
| 25 | 1.86 | 2.66 | 1.75 | 2.59 | 1.65 | 2.53 |
| 30 | 2.61 | 3.19 | 2.45 | 3.11 | 2.31 | 3.04 |
| 35 | 3.47 | 3.73 | 3.26 | 3.63 | 3.07 | 3.54 |
| 40 | 4.45 | 4.26 | 4.17 | 4.15 | 3.93 | 4.05 |
| 45 | 5.53 | 4.79 | 5.18 | 4.67 | 4.89 | 4.55 |
| 50 | 6.72 | 5.32 | 6.30 | 5.18 | 5.94 | 5.06 |
| 60 | 9.42 | 6.39 | 8.83 | 6.22 | 8.33 | 6.07 |
| 70 | 12.53 | 7.45 | 11.75 | 7.26 | 11.08 | 7.08 |
| 80 | 16.05 | 8.52 | 15.05 | 8.29 | 14.19 | 8.10 |
| 90 | 19.96 | 9.58 | 18.72 | 9.33 | 17.65 | 9.11 |
| 100 | 24.26 | 10.64 | 22.75 | 10.37 | 21.45 | 10.12 |

*ASTM B88

**Table B-6** Copper water tube properties—C=130—2½ in. nominal diameter

| *C* = 130 | ID = Inside diameter* | | | | | |
|---|---|---|---|---|---|---|
| | Type K<br>ID = 2.435 | | Type L<br>ID = 2.465 | | Type M<br>ID = 2.495 | |
| Flow<br>*gpm* | Head Loss<br>*ft/100 ft* | Velocity<br>*ft/s* | Head Loss<br>*ft/100 ft* | Velocity<br>*ft/s* | Head Loss<br>*ft/100 ft* | Velocity<br>*ft/s* |
| 20 | 0.43 | 1.38 | 0.40 | 1.34 | 0.38 | 1.31 |
| 25 | 0.65 | 1.72 | 0.61 | 1.68 | 0.57 | 1.64 |
| 30 | 0.90 | 2.07 | 0.85 | 2.02 | 0.80 | 1.97 |
| 35 | 1.20 | 2.41 | 1.13 | 2.35 | 1.07 | 2.30 |
| 40 | 1.54 | 2.76 | 1.45 | 2.69 | 1.37 | 2.62 |
| 45 | 1.92 | 3.10 | 1.81 | 3.03 | 1.70 | 2.95 |
| 50 | 2.33 | 3.44 | 2.19 | 3.36 | 2.07 | 3.28 |
| 60 | 3.27 | 4.13 | 3.08 | 4.03 | 2.90 | 3.94 |
| 70 | 4.34 | 4.82 | 4.09 | 4.71 | 3.86 | 4.59 |
| 80 | 5.56 | 5.51 | 5.24 | 5.38 | 4.94 | 5.25 |
| 90 | 6.92 | 6.20 | 6.52 | 6.05 | 6.14 | 5.91 |
| 100 | 8.41 | 6.89 | 7.92 | 6.72 | 7.47 | 6.56 |
| 110 | 10.03 | 7.58 | 9.45 | 7.40 | 8.91 | 7.22 |
| 120 | 11.79 | 8.27 | 11.10 | 8.07 | 10.47 | 7.87 |
| 130 | 13.67 | 8.96 | 12.88 | 8.74 | 12.14 | 8.53 |
| 140 | 15.68 | 9.65 | 14.77 | 9.41 | 13.93 | 9.19 |
| 150 | 17.82 | 10.33 | 16.79 | 10.08 | 15.83 | 9.84 |
| 160 | 20.08 | 11.02 | 18.92 | 10.76 | 17.83 | 10.50 |

*ASTM B88

**Table B-7** Copper water tube properties—C=130—3 in. nominal diameter

| *C* = 130 | ID = Inside diameter* | | | | | |
|---|---|---|---|---|---|---|
| | Type K<br>ID = 2.907 | | Type L<br>ID = 2.945 | | Type M<br>ID = 2.981 | |
| Flow<br>*gpm* | Head Loss<br>*ft/100 ft* | Velocity<br>*ft/s* | Head Loss<br>*ft/100 ft* | Velocity<br>*ft/s* | Head Loss<br>*ft/100 ft* | Velocity<br>*ft/s* |
| 20 | 0.18 | 0.97 | 0.17 | 0.94 | 0.16 | 0.92 |
| 30 | 0.38 | 1.45 | 0.36 | 1.41 | 0.34 | 1.38 |
| 40 | 0.65 | 1.93 | 0.61 | 1.88 | 0.58 | 1.84 |
| 50 | 0.98 | 2.42 | 0.92 | 2.35 | 0.87 | 2.30 |
| 60 | 1.38 | 2.90 | 1.29 | 2.83 | 1.22 | 2.76 |
| 70 | 1.83 | 3.38 | 1.72 | 3.30 | 1.62 | 3.22 |
| 80 | 2.35 | 3.87 | 2.20 | 3.77 | 2.08 | 3.68 |
| 90 | 2.92 | 4.35 | 2.74 | 4.24 | 2.58 | 4.14 |
| 100 | 3.55 | 4.83 | 3.33 | 4.71 | 3.14 | 4.60 |
| 110 | 4.23 | 5.32 | 3.97 | 5.18 | 3.75 | 5.06 |
| 120 | 4.97 | 5.80 | 4.67 | 5.65 | 4.40 | 5.52 |
| 130 | 5.77 | 6.28 | 5.41 | 6.12 | 5.10 | 5.98 |
| 140 | 6.62 | 6.77 | 6.21 | 6.59 | 5.85 | 6.44 |
| 150 | 7.52 | 7.25 | 7.06 | 7.06 | 6.65 | 6.90 |
| 160 | 8.47 | 7.73 | 7.95 | 7.54 | 7.50 | 7.36 |
| 170 | 9.48 | 8.22 | 8.90 | 8.01 | 8.39 | 7.81 |
| 180 | 10.54 | 8.70 | 9.89 | 8.48 | 9.32 | 8.27 |
| 190 | 11.65 | 9.18 | 10.93 | 8.95 | 10.30 | 8.73 |
| 200 | 12.81 | 9.67 | 12.02 | 9.42 | 11.33 | 9.19 |
| 210 | 14.02 | 10.15 | 13.16 | 9.89 | 12.40 | 9.65 |
| 220 | 15.28 | 10.63 | 14.34 | 10.36 | 13.52 | 10.11 |

*ASTM B88

**Table B-8** Copper water tube properties—C=130—3½ in. nominal diameter

| *C* = 130 | ID = Inside diameter* | | | | | |
|---|---|---|---|---|---|---|
| | Type K<br>ID = 3.385 | | Type L<br>ID = 3.425 | | Type M<br>ID = 3.459 | |
| Flow<br>*gpm* | Head Loss<br>*ft/100 ft* | Velocity<br>*ft/s* | Head Loss<br>*ft/100 ft* | Velocity<br>*ft/s* | Head Loss<br>*ft/100 ft* | Velocity<br>*ft/s* |
| 60 | 0.66 | 2.14 | 0.62 | 2.09 | 0.59 | 2.05 |
| 70 | 0.87 | 2.50 | 0.82 | 2.44 | 0.79 | 2.39 |
| 80 | 1.12 | 2.85 | 1.06 | 2.79 | 1.01 | 2.73 |
| 90 | 1.39 | 3.21 | 1.31 | 3.13 | 1.25 | 3.07 |
| 100 | 1.69 | 3.57 | 1.60 | 3.48 | 1.52 | 3.41 |
| 110 | 2.02 | 3.92 | 1.90 | 3.83 | 1.82 | 3.76 |
| 120 | 2.37 | 4.28 | 2.24 | 4.18 | 2.13 | 4.10 |
| 130 | 2.75 | 4.63 | 2.60 | 4.53 | 2.47 | 4.44 |
| 140 | 3.15 | 4.99 | 2.98 | 4.88 | 2.84 | 4.78 |

*(continued)*

**Table B-8** Copper water tube properties—C=130—3½ in. nominal diameter (*Continued*)

| *C* = 130 | ID = Inside diameter* | | | | | |
|---|---|---|---|---|---|---|
| | Type K ID = 3.385 | | Type L ID = 3.425 | | Type M ID = 3.459 | |
| Flow *gpm* | Head Loss *ft/100 ft* | Velocity *ft/s* | Head Loss *ft/100 ft* | Velocity *ft/s* | Head Loss *ft/100 ft* | Velocity *ft/s* |
| 150 | 3.58 | 5.35 | 3.38 | 5.22 | 3.22 | 5.12 |
| 160 | 4.04 | 5.70 | 3.81 | 5.57 | 3.63 | 5.46 |
| 170 | 4.52 | 6.06 | 4.26 | 5.92 | 4.06 | 5.80 |
| 180 | 5.02 | 6.42 | 4.74 | 6.27 | 4.52 | 6.15 |
| 190 | 5.55 | 6.77 | 5.24 | 6.62 | 4.99 | 6.49 |
| 200 | 6.10 | 7.13 | 5.76 | 6.96 | 5.49 | 6.83 |
| 220 | 7.28 | 7.84 | 6.87 | 7.66 | 6.55 | 7.51 |
| 240 | 8.55 | 8.56 | 8.08 | 8.36 | 7.70 | 8.19 |
| 260 | 9.92 | 9.27 | 9.37 | 9.05 | 8.93 | 8.88 |
| 280 | 11.38 | 9.98 | 10.75 | 9.75 | 10.24 | 9.56 |
| 300 | 12.93 | 10.70 | 12.21 | 10.45 | 11.64 | 10.24 |

*ASTM B88

**Table B-9** Copper water tube properties—C=130—4 in. nominal diameter

| *C* = 130 | ID = Inside diameter* | | | | | |
|---|---|---|---|---|---|---|
| | Type K ID = 3.857 | | Type L ID = 3.905 | | Type M ID = 3.935 | |
| Flow *gpm* | Head Loss *ft/100 ft* | Velocity *ft/s* | Head Loss *ft/100 ft* | Velocity *ft/s* | Head Loss *ft/100 ft* | Velocity *ft/s* |
| 100 | 0.90 | 2.75 | 0.84 | 2.68 | 0.81 | 2.64 |
| 110 | 1.07 | 3.02 | 1.01 | 2.95 | 0.97 | 2.90 |
| 120 | 1.25 | 3.30 | 1.18 | 3.21 | 1.14 | 3.17 |
| 130 | 1.46 | 3.57 | 1.37 | 3.48 | 1.32 | 3.43 |
| 140 | 1.67 | 3.84 | 1.57 | 3.75 | 1.51 | 3.69 |
| 150 | 1.90 | 4.12 | 1.79 | 4.02 | 1.72 | 3.96 |
| 160 | 2.14 | 4.39 | 2.01 | 4.29 | 1.94 | 4.22 |
| 170 | 2.39 | 4.67 | 2.25 | 4.55 | 2.17 | 4.48 |
| 180 | 2.66 | 4.94 | 2.50 | 4.82 | 2.41 | 4.75 |
| 190 | 2.94 | 5.22 | 2.77 | 5.09 | 2.67 | 5.01 |
| 200 | 3.23 | 5.49 | 3.04 | 5.36 | 2.93 | 5.28 |
| 220 | 3.85 | 6.04 | 3.63 | 5.89 | 3.50 | 5.80 |
| 240 | 4.53 | 6.59 | 4.26 | 6.43 | 4.11 | 6.33 |
| 260 | 5.25 | 7.14 | 4.95 | 6.97 | 4.76 | 6.86 |
| 280 | 6.02 | 7.69 | 5.67 | 7.50 | 5.47 | 7.39 |
| 300 | 6.85 | 8.24 | 6.45 | 8.04 | 6.21 | 7.91 |
| 350 | 9.11 | 9.61 | 8.58 | 9.38 | 8.26 | 9.23 |
| 400 | 11.66 | 10.98 | 10.98 | 10.72 | 10.58 | 10.55 |

*ASTM B88

**Table B-10** Copper water tube properties—C=130—5 in. nominal diameter

| *C* = 130 | ID = Inside diameter* | | | | | |
|---|---|---|---|---|---|---|
| | Type K ID = 4.805 | | Type L ID = 4.875 | | Type M ID = 4.907 | |
| Flow *gpm* | Head Loss *ft/100 ft* | Velocity *ft/s* | Head Loss *ft/100 ft* | Velocity *ft/s* | Head Loss *ft/100 ft* | Velocity *ft/s* |
| 150 | 0.65 | 2.65 | 0.61 | 2.58 | 0.59 | 2.54 |
| 160 | 0.73 | 2.83 | 0.68 | 2.75 | 0.66 | 2.71 |
| 170 | 0.82 | 3.01 | 0.76 | 2.92 | 0.74 | 2.88 |
| 180 | 0.91 | 3.18 | 0.85 | 3.09 | 0.82 | 3.05 |
| 190 | 1.01 | 3.36 | 0.94 | 3.27 | 0.91 | 3.22 |
| 200 | 1.11 | 3.54 | 1.03 | 3.44 | 1.00 | 3.39 |
| 220 | 1.32 | 3.89 | 1.23 | 3.78 | 1.19 | 3.73 |
| 240 | 1.55 | 4.25 | 1.45 | 4.13 | 1.40 | 4.07 |
| 260 | 1.80 | 4.60 | 1.68 | 4.47 | 1.63 | 4.41 |
| 280 | 2.07 | 4.95 | 1.93 | 4.81 | 1.86 | 4.75 |
| 300 | 2.35 | 5.31 | 2.19 | 5.16 | 2.12 | 5.09 |
| 350 | 3.12 | 6.19 | 2.91 | 6.02 | 2.82 | 5.94 |
| 400 | 4.00 | 7.08 | 3.73 | 6.88 | 3.61 | 6.79 |
| 450 | 4.97 | 7.96 | 4.64 | 7.73 | 4.49 | 7.63 |
| 500 | 6.05 | 8.85 | 5.63 | 8.59 | 5.46 | 8.48 |
| 550 | 7.21 | 9.73 | 6.72 | 9.45 | 6.51 | 9.33 |
| 600 | 8.47 | 10.62 | 7.90 | 10.31 | 7.65 | 10.18 |

*ASTM B88

**Table B-11** Copper water tube properties—C=130—6 in. nominal diameter

| *C* = 130 | ID = Inside diameter* | | | | | |
|---|---|---|---|---|---|---|
| | Type K ID = 5.741 | | Type L ID = 5.845 | | Type M ID = 5.881 | |
| Flow *gpm* | Head Loss *ft/100 ft* | Velocity *ft/s* | Head Loss *ft/100 ft* | Velocity *ft/s* | Head Loss *ft/100 ft* | Velocity *ft/s* |
| 240 | 0.65 | 2.97 | 0.60 | 2.87 | 0.58 | 2.83 |
| 260 | 0.76 | 3.22 | 0.69 | 3.11 | 0.67 | 3.07 |
| 280 | 0.87 | 3.47 | 0.80 | 3.35 | 0.77 | 3.31 |
| 300 | 0.99 | 3.72 | 0.90 | 3.59 | 0.88 | 3.54 |
| 350 | 1.31 | 4.34 | 1.20 | 4.18 | 1.17 | 4.13 |
| 400 | 1.68 | 4.96 | 1.54 | 4.78 | 1.49 | 4.72 |
| 450 | 2.09 | 5.58 | 1.92 | 5.38 | 1.86 | 5.31 |
| 500 | 2.54 | 6.20 | 2.33 | 5.98 | 2.26 | 5.91 |
| 550 | 3.03 | 6.82 | 2.78 | 6.58 | 2.70 | 6.50 |
| 600 | 3.56 | 7.44 | 3.26 | 7.17 | 3.17 | 7.09 |
| 650 | 4.13 | 8.06 | 3.78 | 7.77 | 3.67 | 7.68 |
| 700 | 4.74 | 8.68 | 4.34 | 8.37 | 4.21 | 8.27 |
| 750 | 5.38 | 9.30 | 4.93 | 8.97 | 4.79 | 8.86 |
| 800 | 6.07 | 9.92 | 5.56 | 9.57 | 5.40 | 9.45 |
| 850 | 6.79 | 10.53 | 6.22 | 10.16 | 6.04 | 10.04 |

*ASTM B88

**Table B-12** Copper water tube properties—C=130—8 in. nominal diameter

| *C* = 130 | ID = Inside diameter* | | | | | |
|---|---|---|---|---|---|---|
| | Type K<br>ID = 7.583 | | Type L<br>ID = 7.725 | | Type M<br>ID = 7.785 | |
| Flow<br>*gpm* | Head Loss<br>*ft/100 ft* | Velocity<br>*ft/s* | Head Loss<br>*ft/100 ft* | Velocity<br>*ft/s* | Head Loss<br>*ft/100 ft* | Velocity<br>*ft/s* |
| 500 | 0.66 | 3.55 | 0.60 | 3.42 | 0.58 | 3.37 |
| 550 | 0.78 | 3.91 | 0.71 | 3.76 | 0.69 | 3.71 |
| 600 | 0.92 | 4.26 | 0.84 | 4.11 | 0.81 | 4.04 |
| 650 | 1.07 | 4.62 | 0.97 | 4.45 | 0.94 | 4.38 |
| 700 | 1.22 | 4.97 | 1.12 | 4.79 | 1.07 | 4.72 |
| 750 | 1.39 | 5.33 | 1.27 | 5.13 | 1.22 | 5.06 |
| 800 | 1.56 | 5.68 | 1.43 | 5.48 | 1.38 | 5.39 |
| 850 | 1.75 | 6.04 | 1.60 | 5.82 | 1.54 | 5.73 |
| 900 | 1.95 | 6.39 | 1.78 | 6.16 | 1.71 | 6.07 |
| 950 | 2.15 | 6.75 | 1.96 | 6.50 | 1.89 | 6.40 |
| 1000 | 2.37 | 7.10 | 2.16 | 6.85 | 2.08 | 6.74 |
| 1100 | 2.82 | 7.81 | 2.58 | 7.53 | 2.48 | 7.41 |
| 1200 | 3.31 | 8.52 | 3.03 | 8.21 | 2.92 | 8.09 |
| 1300 | 3.84 | 9.24 | 3.51 | 8.90 | 3.38 | 8.76 |
| 1400 | 4.41 | 9.95 | 4.03 | 9.58 | 3.88 | 9.44 |
| 1500 | 5.01 | 10.66 | 4.58 | 10.27 | 4.41 | 10.11 |

*ASTM B88

**Table B-13** Polyvinyl chloride (PVC) pipe—C=130—¾ in. nominal diameter

| *C* = 130 | ID = Inside diameter* | | | | | |
|---|---|---|---|---|---|---|
| | SDR17<br>ID = 0.886 | | SDR21<br>ID = 0.89 | | SDR26<br>ID = N/A† | |
| Flow<br>*gpm* | Head Loss<br>*ft/100 ft* | Velocity<br>*ft/s* | Head Loss<br>*ft/100 ft* | Velocity<br>*ft/s* | Head Loss<br>*ft/100 ft* | Velocity<br>*ft/s* |
| 1 | 0.23 | 0.52 | 0.22 | 0.52 | N/A | N/A |
| 2 | 0.83 | 1.04 | 0.81 | 1.03 | N/A | N/A |
| 3 | 1.75 | 1.56 | 1.71 | 1.55 | N/A | N/A |
| 4 | 2.98 | 2.08 | 2.92 | 2.06 | N/A | N/A |
| 5 | 4.51 | 2.60 | 4.41 | 2.58 | N/A | N/A |
| 6 | 6.32 | 3.12 | 6.18 | 3.09 | N/A | N/A |
| 7 | 8.40 | 3.64 | 8.22 | 3.61 | N/A | N/A |
| 8 | 10.76 | 4.16 | 10.53 | 4.13 | N/A | N/A |
| 9 | 13.38 | 4.68 | 13.09 | 4.64 | N/A | N/A |
| 10 | 16.27 | 5.20 | 15.91 | 5.16 | N/A | N/A |
| 11 | 19.41 | 5.72 | 18.99 | 5.67 | N/A | N/A |
| 12 | 22.80 | 6.24 | 22.31 | 6.19 | N/A | N/A |
| 13 | 26.44 | 6.76 | 25.87 | 6.70 | N/A | N/A |
| 14 | 30.33 | 7.29 | 29.67 | 7.22 | N/A | N/A |

*(continued)*

## Table B-13 Polyvinyl chloride (PVC) pipe—C=130—¾ in. nominal diameter (*Continued*)

| C = 130 | ID = Inside diameter* | | | | | |
|---|---|---|---|---|---|---|
| | SDR17 ID = 0.886 | | SDR21 ID = 0.89 | | SDR26 ID = N/A† | |
| Flow *gpm* | Head Loss *ft/100 ft* | Velocity *ft/s* | Head Loss *ft/100 ft* | Velocity *ft/s* | Head Loss *ft/100 ft* | Velocity *ft/s* |
| 15 | 34.47 | 7.81 | 33.72 | 7.74 | N/A | N/A |
| 16 | 38.84 | 8.33 | 38.00 | 8.25 | N/A | N/A |
| 17 | 43.46 | 8.85 | 42.51 | 8.77 | N/A | N/A |
| 18 | 48.31 | 9.37 | 47.26 | 9.28 | N/A | N/A |
| 19 | 53.40 | 9.89 | 52.24 | 9.80 | N/A | N/A |
| 20 | 58.72 | 10.41 | 57.44 | 10.31 | N/A | N/A |

*ASTM D2241
†NA = not applicable

## Table B-14 Polyvinyl chloride (PVC) pipe—C=130—1 in. nominal diameter

| C = 130 | ID = Inside diameter* | | | | | |
|---|---|---|---|---|---|---|
| | SDR17 ID = 1.121 | | SDR21 ID = 1.149 | | SDR26 ID = 1.155 | |
| Flow *gpm* | Head Loss *ft/100 ft* | Velocity *ft/s* | Head Loss *ft/100 ft* | Velocity *ft/s* | Head Loss *ft/100 ft* | Velocity *ft/s* |
| 1 | 0.07 | 0.33 | 0.06 | 0.31 | 0.06 | 0.31 |
| 2 | 0.26 | 0.65 | 0.23 | 0.62 | 0.23 | 0.61 |
| 3 | 0.56 | 0.98 | 0.49 | 0.93 | 0.48 | 0.92 |
| 4 | 0.95 | 1.30 | 0.84 | 1.24 | 0.82 | 1.22 |
| 5 | 1.43 | 1.63 | 1.27 | 1.55 | 1.24 | 1.53 |
| 6 | 2.01 | 1.95 | 1.78 | 1.86 | 1.74 | 1.84 |
| 7 | 2.67 | 2.28 | 2.37 | 2.17 | 2.31 | 2.14 |
| 8 | 3.42 | 2.60 | 3.03 | 2.48 | 2.96 | 2.45 |
| 9 | 4.26 | 2.93 | 3.77 | 2.78 | 3.68 | 2.76 |
| 10 | 5.17 | 3.25 | 4.59 | 3.09 | 4.47 | 3.06 |
| 12 | 7.25 | 3.90 | 6.43 | 3.71 | 6.27 | 3.67 |
| 14 | 9.64 | 4.55 | 8.55 | 4.33 | 8.34 | 4.29 |
| 16 | 12.35 | 5.20 | 10.95 | 4.95 | 10.68 | 4.90 |
| 18 | 15.36 | 5.85 | 13.62 | 5.57 | 13.28 | 5.51 |
| 20 | 18.67 | 6.50 | 16.56 | 6.19 | 16.14 | 6.12 |
| 25 | 28.22 | 8.13 | 25.03 | 7.74 | 24.40 | 7.66 |
| 30 | 39.56 | 9.75 | 35.08 | 9.28 | 34.20 | 9.19 |
| 35 | 52.63 | 11.38 | 46.67 | 10.83 | 45.50 | 10.72 |

*ASTM D2241

**Table B-15** Polyvinyl chloride (PVC) pipe—C=130—1¼ in. nominal diameter

| *C* = 130 | ID = Inside diameter* | | | | | |
|---|---|---|---|---|---|---|
| | SDR17 ID = 1.424 | | SDR21 ID = 1.462 | | SDR26 ID = 1.492 | |
| Flow *gpm* | Head Loss *ft/100 ft* | Velocity *ft/s* | Head Loss *ft/100 ft* | Velocity *ft/s* | Head Loss *ft/100 ft* | Velocity *ft/s* |
| 5 | 0.45 | 1.01 | 0.39 | 0.96 | 0.36 | 0.92 |
| 6 | 0.63 | 1.21 | 0.55 | 1.15 | 0.50 | 1.10 |
| 7 | 0.83 | 1.41 | 0.73 | 1.34 | 0.66 | 1.28 |
| 8 | 1.07 | 1.61 | 0.94 | 1.53 | 0.85 | 1.47 |
| 9 | 1.33 | 1.81 | 1.17 | 1.72 | 1.06 | 1.65 |
| 10 | 1.61 | 2.01 | 1.42 | 1.91 | 1.29 | 1.84 |
| 12 | 2.26 | 2.42 | 1.99 | 2.29 | 1.80 | 2.20 |
| 15 | 3.42 | 3.02 | 3.01 | 2.87 | 2.72 | 2.75 |
| 20 | 5.82 | 4.03 | 5.12 | 3.82 | 4.64 | 3.67 |
| 25 | 8.80 | 5.04 | 7.74 | 4.78 | 7.01 | 4.59 |
| 30 | 12.34 | 6.04 | 10.85 | 5.73 | 9.83 | 5.51 |
| 35 | 16.41 | 7.05 | 14.44 | 6.69 | 13.08 | 6.42 |
| 40 | 21.02 | 8.06 | 18.49 | 7.64 | 16.75 | 7.34 |
| 45 | 26.14 | 9.07 | 22.99 | 8.60 | 20.83 | 8.26 |
| 50 | 31.77 | 10.07 | 27.95 | 9.56 | 25.31 | 9.18 |
| 55 | | | 33.34 | 10.51 | 30.20 | 10.09 |

*ASTM D2241

**Table B-16** Polyvinyl chloride (PVC) pipe—C=130—1½ in. nominal diameter

| *C* = 130 | ID = Inside diameter* | | | | | |
|---|---|---|---|---|---|---|
| | SDR17 ID = 1.636 | | SDR21 ID = 1.68 | | SDR26 ID = 1.714 | |
| Flow *gpm* | Head Loss *ft/100 ft* | Velocity *ft/s* | Head Loss *ft/100 ft* | Velocity *ft/s* | Head Loss *ft/100 ft* | Velocity *ft/s* |
| 8 | 0.54 | 1.22 | 0.48 | 1.16 | 0.43 | 1.11 |
| 9 | 0.68 | 1.37 | 0.59 | 1.30 | 0.54 | 1.25 |
| 10 | 0.82 | 1.53 | 0.72 | 1.45 | 0.65 | 1.39 |
| 12 | 1.15 | 1.83 | 1.01 | 1.74 | 0.92 | 1.67 |
| 15 | 1.74 | 2.29 | 1.53 | 2.17 | 1.39 | 2.09 |
| 20 | 2.96 | 3.05 | 2.60 | 2.89 | 2.36 | 2.78 |
| 25 | 4.48 | 3.82 | 3.93 | 3.62 | 3.57 | 3.48 |
| 30 | 6.28 | 4.58 | 5.51 | 4.34 | 5.00 | 4.17 |
| 35 | 8.35 | 5.34 | 7.34 | 5.07 | 6.65 | 4.87 |
| 40 | 10.69 | 6.10 | 9.39 | 5.79 | 8.52 | 5.56 |
| 50 | 16.16 | 7.63 | 14.20 | 7.24 | 12.88 | 6.95 |
| 60 | 22.65 | 9.16 | 19.91 | 8.68 | 18.05 | 8.34 |
| 70 | 30.14 | 10.68 | 26.48 | 10.13 | 24.02 | 9.73 |
| 80 | | | | | 30.76 | 11.12 |

*ASTM D2241

**Table B-17** Polyvinyl chloride (PVC) pipe—C=130—2 in. nominal diameter

| *C* = 130 | ID = Inside diameter* | | | | | |
|---|---|---|---|---|---|---|
| | SDR17 ID = 2.055 | | SDR21 ID = 2.109 | | SDR26 ID = 2.153 | |
| Flow *gpm* | Head Loss *ft/100 ft* | Velocity *ft/s* | Head Loss *ft/100 ft* | Velocity *ft/s* | Head Loss *ft/100 ft* | Velocity *ft/s* |
| 10 | 0.27 | 0.97 | 0.24 | 0.92 | 0.22 | 0.88 |
| 12 | 0.38 | 1.16 | 0.33 | 1.10 | 0.30 | 1.06 |
| 14 | 0.50 | 1.35 | 0.44 | 1.29 | 0.40 | 1.23 |
| 16 | 0.65 | 1.55 | 0.57 | 1.47 | 0.51 | 1.41 |
| 18 | 0.80 | 1.74 | 0.71 | 1.65 | 0.64 | 1.59 |
| 20 | 0.98 | 1.93 | 0.86 | 1.84 | 0.78 | 1.76 |
| 25 | 1.47 | 2.42 | 1.30 | 2.30 | 1.18 | 2.20 |
| 30 | 2.07 | 2.90 | 1.82 | 2.76 | 1.65 | 2.64 |
| 35 | 2.75 | 3.39 | 2.42 | 3.21 | 2.19 | 3.08 |
| 40 | 3.52 | 3.87 | 3.10 | 3.67 | 2.81 | 3.53 |
| 45 | 4.38 | 4.35 | 3.86 | 4.13 | 3.49 | 3.97 |
| 50 | 5.32 | 4.84 | 4.69 | 4.59 | 4.24 | 4.41 |
| 60 | 7.46 | 5.80 | 6.58 | 5.51 | 5.95 | 5.29 |
| 70 | 9.93 | 6.77 | 8.75 | 6.43 | 7.91 | 6.17 |
| 80 | 12.71 | 7.74 | 11.20 | 7.35 | 10.13 | 7.05 |
| 90 | 15.81 | 8.71 | 13.93 | 8.27 | 12.60 | 7.93 |
| 100 | 19.21 | 9.67 | 16.93 | 9.18 | 15.31 | 8.81 |
| 110 | 22.92 | 10.64 | 20.20 | 10.10 | 18.27 | 9.69 |
| 120 | | | | | 21.46 | 10.58 |

*ASTM D2241

**Table B-18** Polyvinyl chloride (PVC) pipe—C=130—2½ in. nominal diameter

| *C* = 130 | ID = Inside diameter* | | | | | |
|---|---|---|---|---|---|---|
| | SDR17 ID = 2.497 | | SDR21 ID = 2.561 | | SDR26 ID = 2.615 | |
| Flow *gpm* | Head Loss *ft/100 ft* | Velocity *ft/s* | Head Loss *ft/100 ft* | Velocity *ft/s* | Head Loss *ft/100 ft* | Velocity *ft/s* |
| 20 | 0.38 | 1.31 | 0.33 | 1.25 | 0.30 | 1.19 |
| 25 | 0.57 | 1.64 | 0.50 | 1.56 | 0.46 | 1.49 |
| 30 | 0.80 | 1.97 | 0.71 | 1.87 | 0.64 | 1.79 |
| 35 | 1.06 | 2.29 | 0.94 | 2.18 | 0.85 | 2.09 |
| 40 | 1.36 | 2.62 | 1.21 | 2.49 | 1.09 | 2.39 |
| 45 | 1.70 | 2.95 | 1.50 | 2.80 | 1.35 | 2.69 |
| 50 | 2.06 | 3.28 | 1.82 | 3.11 | 1.65 | 2.99 |
| 60 | 2.89 | 3.93 | 2.55 | 3.74 | 2.31 | 3.58 |
| 70 | 3.84 | 4.59 | 3.40 | 4.36 | 3.07 | 4.18 |
| 80 | 4.92 | 5.24 | 4.35 | 4.98 | 3.93 | 4.78 |
| 90 | 6.12 | 5.90 | 5.41 | 5.61 | 4.89 | 5.38 |

*(continued)*

**Table B-18** Polyvinyl chloride (PVC) pipe—C=130—2½ in. nominal diameter (*Continued*)

| *C* = 130 | ID = Inside diameter* | | | | | |
|---|---|---|---|---|---|---|
| | SDR17 ID = 2.497 | | SDR21 ID = 2.561 | | SDR26 ID = 2.615 | |
| **Flow** *gpm* | **Head Loss** *ft/100 ft* | **Velocity** *ft/s* | **Head Loss** *ft/100 ft* | **Velocity** *ft/s* | **Head Loss** *ft/100 ft* | **Velocity** *ft/s* |
| 100 | 7.44 | 6.55 | 6.58 | 6.23 | 5.94 | 5.97 |
| 110 | 8.88 | 7.21 | 7.85 | 6.85 | 7.09 | 6.57 |
| 120 | 10.43 | 7.86 | 9.22 | 7.47 | 8.33 | 7.17 |
| 130 | 12.09 | 8.52 | 10.69 | 8.10 | 9.66 | 7.77 |
| 140 | 13.87 | 9.17 | 12.26 | 8.72 | 11.08 | 8.36 |
| 150 | 15.76 | 9.83 | 13.94 | 9.34 | 12.59 | 8.96 |
| 160 | 17.76 | 10.48 | 15.70 | 9.97 | 14.19 | 9.56 |
| 170 | | | 17.57 | 10.59 | 15.87 | 10.16 |

*ASTM D2241

**Table B-19** Polyvinyl chloride (PVC) pipe—C=130—3 in. nominal diameter

| *C* = 130 | ID = Inside diameter* | | | | | |
|---|---|---|---|---|---|---|
| | SDR17 ID = 3.038 | | SDR21 ID = 3.126 | | SDR26 ID = 3.19 | |
| **Flow** *gpm* | **Head Loss** *ft/100 ft* | **Velocity** *ft/s* | **Head Loss** *ft/100 ft* | **Velocity** *ft/s* | **Head Loss** *ft/100 ft* | **Velocity** *ft/s* |
| 20 | 0.15 | 0.89 | 0.13 | 0.84 | 0.11 | 0.80 |
| 30 | 0.31 | 1.33 | 0.27 | 1.25 | 0.24 | 1.20 |
| 40 | 0.52 | 1.77 | 0.46 | 1.67 | 0.41 | 1.61 |
| 50 | 0.79 | 2.21 | 0.69 | 2.09 | 0.63 | 2.01 |
| 60 | 1.11 | 2.66 | 0.97 | 2.51 | 0.88 | 2.41 |
| 70 | 1.48 | 3.10 | 1.29 | 2.93 | 1.17 | 2.81 |
| 80 | 1.89 | 3.54 | 1.65 | 3.34 | 1.49 | 3.21 |
| 90 | 2.36 | 3.98 | 2.05 | 3.76 | 1.86 | 3.61 |
| 100 | 2.86 | 4.43 | 2.49 | 4.18 | 2.26 | 4.01 |
| 110 | 3.42 | 4.87 | 2.97 | 4.60 | 2.69 | 4.42 |
| 120 | 4.01 | 5.31 | 3.49 | 5.02 | 3.16 | 4.82 |
| 130 | 4.65 | 5.75 | 4.05 | 5.43 | 3.67 | 5.22 |
| 140 | 5.34 | 6.20 | 4.64 | 5.85 | 4.21 | 5.62 |
| 150 | 6.07 | 6.64 | 5.28 | 6.27 | 4.78 | 6.02 |
| 160 | 6.84 | 7.08 | 5.95 | 6.69 | 5.39 | 6.42 |
| 170 | 7.65 | 7.52 | 6.65 | 7.11 | 6.03 | 6.82 |
| 180 | 8.50 | 7.97 | 7.40 | 7.52 | 6.70 | 7.23 |
| 190 | 9.40 | 8.41 | 8.18 | 7.94 | 7.41 | 7.63 |
| 200 | 10.33 | 8.85 | 8.99 | 8.36 | 8.15 | 8.03 |
| 210 | 11.31 | 9.29 | 9.84 | 8.78 | 8.92 | 8.43 |
| 220 | 12.33 | 9.74 | 10.73 | 9.20 | 9.72 | 8.83 |
| 230 | 13.39 | 10.18 | 11.65 | 9.61 | 10.55 | 9.23 |
| 240 | | | 12.60 | 10.03 | 11.42 | 9.63 |
| 250 | | | | | 12.31 | 10.04 |

*ASTM D2241

**Table B-20** Polyvinyl chloride (PVC) pipe—C=130—3½ in. nominal diameter

| C = 130 | ID = Inside diameter* | | | | | |
|---|---|---|---|---|---|---|
| | SDR17 ID = 3.474 | | SDR21 ID = 3.574 | | SDR26 ID = 3.652 | |
| Flow *gpm* | Head Loss *ft/100 ft* | Velocity *ft/s* | Head Loss *ft/100 ft* | Velocity *ft/s* | Head Loss *ft/100 ft* | Velocity *ft/s* |
| 60 | 0.58 | 2.03 | 0.50 | 1.92 | 0.45 | 1.84 |
| 70 | 0.77 | 2.37 | 0.67 | 2.24 | 0.60 | 2.14 |
| 80 | 0.99 | 2.71 | 0.86 | 2.56 | 0.77 | 2.45 |
| 90 | 1.23 | 3.05 | 1.07 | 2.88 | 0.96 | 2.76 |
| 100 | 1.49 | 3.38 | 1.30 | 3.20 | 1.17 | 3.06 |
| 110 | 1.78 | 3.72 | 1.55 | 3.52 | 1.39 | 3.37 |
| 120 | 2.09 | 4.06 | 1.82 | 3.84 | 1.64 | 3.68 |
| 130 | 2.42 | 4.40 | 2.11 | 4.16 | 1.90 | 3.98 |
| 140 | 2.78 | 4.74 | 2.42 | 4.48 | 2.18 | 4.29 |
| 150 | 3.16 | 5.08 | 2.75 | 4.80 | 2.47 | 4.59 |
| 160 | 3.56 | 5.42 | 3.10 | 5.12 | 2.79 | 4.90 |
| 170 | 3.98 | 5.75 | 3.47 | 5.44 | 3.12 | 5.21 |
| 180 | 4.42 | 6.09 | 3.85 | 5.76 | 3.47 | 5.51 |
| 190 | 4.89 | 6.43 | 4.26 | 6.08 | 3.83 | 5.82 |
| 200 | 5.38 | 6.77 | 4.68 | 6.40 | 4.22 | 6.13 |
| 220 | 6.42 | 7.45 | 5.59 | 7.04 | 5.03 | 6.74 |
| 240 | 7.54 | 8.12 | 6.56 | 7.68 | 5.91 | 7.35 |
| 260 | 8.74 | 8.80 | 7.61 | 8.31 | 6.85 | 7.96 |
| 280 | 10.03 | 9.48 | 8.73 | 8.95 | 7.86 | 8.58 |
| 300 | 11.39 | 10.15 | 9.92 | 9.59 | 8.93 | 9.19 |
| 320 | | | 11.18 | 10.23 | 10.07 | 9.80 |
| 340 | | | | | 11.26 | 10.41 |

*ASTM D2241

**Table B-21** Polyvinyl chloride (PVC) pipe—C=130—4 in. nominal diameter

| C = 130 | ID = Inside diameter* | | | | | |
|---|---|---|---|---|---|---|
| | SDR14 ID = 4.114 | | SDR18 ID = 4.266 | | SDR25 ID = 4.116 | |
| Flow *gpm* | Head Loss *ft/100 ft* | Velocity *ft/s* | Head Loss *ft/100 ft* | Velocity *ft/s* | Head Loss *ft/100 ft* | Velocity *ft/s* |
| 100 | 0.65 | 2.41 | 0.55 | 2.24 | 0.65 | 2.41 |
| 110 | 0.78 | 2.65 | 0.65 | 2.47 | 0.78 | 2.65 |
| 120 | 0.92 | 2.90 | 0.77 | 2.69 | 0.91 | 2.89 |
| 130 | 1.06 | 3.14 | 0.89 | 2.92 | 1.06 | 3.13 |
| 140 | 1.22 | 3.38 | 1.02 | 3.14 | 1.22 | 3.38 |
| 150 | 1.39 | 3.62 | 1.16 | 3.37 | 1.38 | 3.62 |
| 160 | 1.56 | 3.86 | 1.31 | 3.59 | 1.56 | 3.86 |
| 170 | 1.75 | 4.10 | 1.46 | 3.82 | 1.74 | 4.10 |

*(continued)*

**Table B-21** Polyvinyl chloride (PVC) pipe—C=130—4 in. nominal diameter (*Continued*)

| *C* = 130 | ID = Inside diameter* | | | | | |
|---|---|---|---|---|---|---|
| | SDR14 ID = 4.114 | | SDR18 ID = 4.266 | | SDR25 ID = 4.116 | |
| Flow *gpm* | Head Loss *ft/100 ft* | Velocity *ft/s* | Head Loss *ft/100 ft* | Velocity *ft/s* | Head Loss *ft/100 ft* | Velocity *ft/s* |
| 180 | 1.94 | 4.34 | 1.63 | 4.04 | 1.94 | 4.34 |
| 190 | 2.15 | 4.59 | 1.80 | 4.26 | 2.14 | 4.58 |
| 200 | 2.36 | 4.83 | 1.98 | 4.49 | 2.35 | 4.82 |
| 220 | 2.82 | 5.31 | 2.36 | 4.94 | 2.81 | 5.30 |
| 240 | 3.31 | 5.79 | 2.77 | 5.39 | 3.30 | 5.79 |
| 260 | 3.84 | 6.28 | 3.21 | 5.84 | 3.83 | 6.27 |
| 280 | 4.40 | 6.76 | 3.69 | 6.29 | 4.39 | 6.75 |
| 300 | 5.00 | 7.24 | 4.19 | 6.73 | 4.99 | 7.23 |
| 350 | 6.65 | 8.45 | 5.57 | 7.86 | 6.64 | 8.44 |
| 400 | 8.52 | 9.65 | 7.14 | 8.98 | 8.50 | 9.64 |
| 450 | | | | | 10.57 | 10.85 |

* ANSI/AWWA C900-97

**Table B-22** Polyvinyl chloride (PVC) pipe—C=130—5 in. nominal diameter

| *C* = 130 | ID = Inside diameter* | | | | | |
|---|---|---|---|---|---|---|
| | SDR17 ID = 4.831 | | SDR21 ID = 4.969 | | SDR26 ID = 5.081 | |
| Flow *gpm* | Head Loss *ft/100 ft* | Velocity *ft/s* | Head Loss *ft/100 ft* | Velocity *ft/s* | Head Loss *ft/100 ft* | Velocity *ft/s* |
| 150 | 0.63 | 2.63 | 0.55 | 2.48 | 0.50 | 2.37 |
| 160 | 0.71 | 2.80 | 0.62 | 2.65 | 0.56 | 2.53 |
| 170 | 0.80 | 2.98 | 0.70 | 2.81 | 0.62 | 2.69 |
| 180 | 0.89 | 3.15 | 0.77 | 2.98 | 0.69 | 2.85 |
| 190 | 0.98 | 3.33 | 0.86 | 3.14 | 0.77 | 3.01 |
| 200 | 1.08 | 3.50 | 0.94 | 3.31 | 0.84 | 3.16 |
| 220 | 1.29 | 3.85 | 1.12 | 3.64 | 1.01 | 3.48 |
| 240 | 1.51 | 4.20 | 1.32 | 3.97 | 1.18 | 3.80 |
| 260 | 1.75 | 4.55 | 1.53 | 4.30 | 1.37 | 4.11 |
| 280 | 2.01 | 4.90 | 1.75 | 4.63 | 1.57 | 4.43 |
| 300 | 2.29 | 5.25 | 1.99 | 4.96 | 1.79 | 4.75 |
| 350 | 3.04 | 6.13 | 2.65 | 5.79 | 2.38 | 5.54 |
| 400 | 3.90 | 7.00 | 3.40 | 6.62 | 3.05 | 6.33 |
| 450 | 4.84 | 7.88 | 4.22 | 7.44 | 3.79 | 7.12 |
| 500 | 5.89 | 8.75 | 5.13 | 8.27 | 4.61 | 7.91 |
| 550 | 7.03 | 9.63 | 6.12 | 9.10 | 5.49 | 8.70 |
| 600 | 8.25 | 10.50 | 7.20 | 9.93 | 6.46 | 9.49 |
| 650 | | | 8.35 | 10.75 | 7.49 | 10.29 |

*ASTM D2241

**Table B-23** Polyvinyl chloride (PVC) pipe—C=130—6 in. nominal diameter

| *C* = 130 | ID = Inside diameter* | | | | | |
|---|---|---|---|---|---|---|
| | SDR14 ID = 5.796 | | SDR18 ID = 6.042 | | SDR25 ID = 6.282 | |
| Flow *gpm* | Head Loss *ft/100 ft* | Velocity *ft/s* | Head Loss *ft/100 ft* | Velocity *ft/s* | Head Loss *ft/100 ft* | Velocity *ft/s* |
| 50 | 0.03 | 0.61 | 0.03 | 0.56 | 0.02 | 0.52 |
| 60 | 0.05 | 0.73 | 0.04 | 0.67 | 0.03 | 0.62 |
| 70 | 0.06 | 0.85 | 0.05 | 0.78 | 0.04 | 0.72 |
| 80 | 0.08 | 0.97 | 0.07 | 0.90 | 0.06 | 0.83 |
| 90 | 0.10 | 1.09 | 0.08 | 1.01 | 0.07 | 0.93 |
| 100 | 0.12 | 1.22 | 0.10 | 1.12 | 0.08 | 1.04 |
| 120 | 0.17 | 1.46 | 0.14 | 1.34 | 0.12 | 1.24 |
| 140 | 0.23 | 1.70 | 0.19 | 1.57 | 0.16 | 1.45 |
| 160 | 0.29 | 1.95 | 0.24 | 1.79 | 0.20 | 1.66 |
| 180 | 0.37 | 2.19 | 0.30 | 2.01 | 0.25 | 1.86 |
| 200 | 0.44 | 2.43 | 0.36 | 2.24 | 0.30 | 2.07 |
| 220 | 0.53 | 2.68 | 0.43 | 2.46 | 0.36 | 2.28 |
| 240 | 0.62 | 2.92 | 0.51 | 2.69 | 0.42 | 2.48 |
| 260 | 0.72 | 3.16 | 0.59 | 2.91 | 0.49 | 2.69 |
| 280 | 0.83 | 3.40 | 0.68 | 3.13 | 0.56 | 2.90 |
| 300 | 0.94 | 3.65 | 0.77 | 3.36 | 0.64 | 3.11 |
| 320 | 1.06 | 3.89 | 0.87 | 3.58 | 0.72 | 3.31 |
| 340 | 1.19 | 4.13 | 0.97 | 3.80 | 0.80 | 3.52 |
| 360 | 1.32 | 4.38 | 1.08 | 4.03 | 0.89 | 3.73 |
| 380 | 1.46 | 4.62 | 1.19 | 4.25 | 0.99 | 3.93 |
| 400 | 1.60 | 4.86 | 1.31 | 4.48 | 1.08 | 4.14 |
| 450 | 2.00 | 5.47 | 1.63 | 5.04 | 1.35 | 4.66 |
| 500 | 2.43 | 6.08 | 1.98 | 5.59 | 1.64 | 5.18 |
| 550 | 2.89 | 6.69 | 2.36 | 6.15 | 1.96 | 5.69 |
| 600 | 3.40 | 7.30 | 2.78 | 6.71 | 2.30 | 6.21 |
| 650 | 3.94 | 7.90 | 3.22 | 7.27 | 2.66 | 6.73 |
| 700 | 4.52 | 8.51 | 3.69 | 7.83 | 3.06 | 7.25 |
| 750 | 5.14 | 9.12 | 4.20 | 8.39 | 3.47 | 7.76 |
| 800 | 5.79 | 9.73 | 4.73 | 8.95 | 3.91 | 8.28 |
| 850 | 6.48 | 10.34 | 5.29 | 9.51 | 4.38 | 8.80 |
| 900 | | | 5.88 | 10.07 | 4.87 | 9.32 |
| 950 | | | | | 5.38 | 9.83 |
| 1000 | | | | | 5.92 | 10.35 |

* ANSI/AWWA C900

**Table B-24** Polyvinyl chloride (PVC) pipe—C=130—8 in. nominal diameter

| *C* = 130 | ID = Inside diameter* | | | | | |
|---|---|---|---|---|---|---|
| | SDR14 ID = 7.602 | | SDR18 ID = 7.924 | | SDR25 ID = 8.24 | |
| **Flow** *gpm* | **Head Loss** *ft/100 ft* | **Velocity** *ft/s* | **Head Loss** *ft/100 ft* | **Velocity** *ft/s* | **Head Loss** *ft/100 ft* | **Velocity** *ft/s* |
| 130 | 0.05 | 0.92 | 0.04 | 0.85 | 0.04 | 0.78 |
| 140 | 0.06 | 0.99 | 0.05 | 0.91 | 0.04 | 0.84 |
| 150 | 0.07 | 1.06 | 0.06 | 0.98 | 0.05 | 0.90 |
| 160 | 0.08 | 1.13 | 0.06 | 1.04 | 0.05 | 0.96 |
| 170 | 0.09 | 1.20 | 0.07 | 1.11 | 0.06 | 1.02 |
| 180 | 0.10 | 1.27 | 0.08 | 1.17 | 0.07 | 1.08 |
| 190 | 0.11 | 1.34 | 0.09 | 1.24 | 0.07 | 1.14 |
| 200 | 0.12 | 1.41 | 0.10 | 1.30 | 0.08 | 1.20 |
| 220 | 0.14 | 1.56 | 0.12 | 1.43 | 0.10 | 1.32 |
| 240 | 0.17 | 1.70 | 0.14 | 1.56 | 0.11 | 1.44 |
| 260 | 0.19 | 1.84 | 0.16 | 1.69 | 0.13 | 1.56 |
| 280 | 0.22 | 1.98 | 0.18 | 1.82 | 0.15 | 1.68 |
| 300 | 0.25 | 2.12 | 0.21 | 1.95 | 0.17 | 1.80 |
| 350 | 0.33 | 2.47 | 0.27 | 2.28 | 0.23 | 2.11 |
| 400 | 0.43 | 2.83 | 0.35 | 2.60 | 0.29 | 2.41 |
| 450 | 0.53 | 3.18 | 0.44 | 2.93 | 0.36 | 2.71 |
| 500 | 0.65 | 3.53 | 0.53 | 3.25 | 0.44 | 3.01 |
| 550 | 0.77 | 3.89 | 0.63 | 3.58 | 0.52 | 3.31 |
| 600 | 0.91 | 4.24 | 0.74 | 3.90 | 0.61 | 3.61 |
| 650 | 1.05 | 4.59 | 0.86 | 4.23 | 0.71 | 3.91 |
| 700 | 1.21 | 4.95 | 0.99 | 4.55 | 0.82 | 4.21 |
| 750 | 1.37 | 5.30 | 1.12 | 4.88 | 0.93 | 4.51 |
| 800 | 1.55 | 5.65 | 1.26 | 5.20 | 1.04 | 4.81 |
| 850 | 1.73 | 6.01 | 1.41 | 5.53 | 1.17 | 5.11 |
| 900 | 1.92 | 6.36 | 1.57 | 5.86 | 1.30 | 5.41 |
| 950 | 2.12 | 6.72 | 1.74 | 6.18 | 1.43 | 5.72 |
| 1000 | 2.34 | 7.07 | 1.91 | 6.51 | 1.58 | 6.02 |
| 1100 | 2.79 | 7.78 | 2.28 | 7.16 | 1.88 | 6.62 |
| 1200 | 3.27 | 8.48 | 2.68 | 7.81 | 2.21 | 7.22 |
| 1300 | 3.80 | 9.19 | 3.10 | 8.46 | 2.57 | 7.82 |
| 1400 | 4.36 | 9.90 | 3.56 | 9.11 | 2.94 | 8.42 |
| 1500 | 4.95 | 10.60 | 4.04 | 9.76 | 3.34 | 9.02 |
| 1600 | | | 4.56 | 10.41 | 3.77 | 9.63 |
| 1700 | | | | | 4.22 | 10.23 |

* ANSI/AWWA C900

**Table B-25** Polyethylene pipe—C=130—¾ in. nominal diameter

| *C* = 130 | ID = Inside diameter* | | | | | | | |
|---|---|---|---|---|---|---|---|---|
| | Copper Tubing Sizes | | | | Iron Pipe Sizes | | | |
| | DR 9 ID = 0.681 | | DR 11 ID = 0.715 | | DR 9 ID = 0.816 | | DR 11 ID = 0.86 | |
| Flow *gpm* | Head Loss *ft/100 ft* | Velocity *ft/s* | Head Loss *ft/100 ft* | Velocity *ft/s* | Head Loss *ft/100 ft* | Velocity *ft/s* | Head Loss *ft/100 ft* | Velocity *ft/s* |
| 1 | 0.82 | 0.88 | 0.65 | 0.80 | 0.34 | 0.61 | 0.26 | 0.55 |
| 2 | 2.98 | 1.76 | 2.35 | 1.60 | 1.23 | 1.23 | 0.95 | 1.10 |
| 3 | 6.30 | 2.64 | 4.97 | 2.40 | 2.61 | 1.84 | 2.02 | 1.66 |
| 4 | 10.74 | 3.52 | 8.47 | 3.20 | 4.45 | 2.45 | 3.45 | 2.21 |
| 5 | 16.24 | 4.40 | 12.81 | 4.00 | 6.73 | 3.07 | 5.21 | 2.76 |
| 6 | 22.76 | 5.29 | 17.95 | 4.79 | 9.43 | 3.68 | 7.30 | 3.31 |
| 7 | 30.27 | 6.17 | 23.88 | 5.59 | 12.55 | 4.29 | 9.72 | 3.87 |
| 8 | 38.77 | 7.05 | 30.58 | 6.39 | 16.07 | 4.91 | 12.44 | 4.42 |
| 9 | 48.22 | 7.93 | 38.03 | 7.19 | 19.98 | 5.52 | 15.47 | 4.97 |
| 10 | 58.60 | 8.81 | 46.22 | 7.99 | 24.29 | 6.13 | 18.81 | 5.52 |
| 11 | 69.92 | 9.69 | 55.15 | 8.79 | 28.98 | 6.75 | 22.44 | 6.08 |
| 12 | 82.14 | 10.57 | 64.79 | 9.59 | 34.04 | 7.36 | 26.36 | 6.63 |
| 13 | | | 75.14 | 10.39 | 39.48 | 7.98 | 30.57 | 7.18 |
| 14 | | | | | 45.29 | 8.59 | 35.07 | 7.73 |
| 15 | | | | | 51.46 | 9.20 | 39.85 | 8.28 |
| 16 | | | | | 57.99 | 9.82 | 44.91 | 8.84 |
| 17 | | | | | 64.89 | 10.43 | 50.24 | 9.39 |
| 18 | | | | | | 0.61 | 55.85 | 9.94 |
| 19 | | | | | | 1.23 | 61.73 | 10.49 |

*ANSI/AWWA C901 (Average outside diameter less minimum wall thickness)

**Table B-26** Polyethylene pipe—C=130—1 in. nominal diameter

| *C* = 130 | ID = Inside diameter* | | | | | | | |
|---|---|---|---|---|---|---|---|---|
| | Copper Tubing Sizes | | | | Iron Pipe Sizes | | | |
| | DR 9 ID = 0.875 | | DR 11 ID = 0.921 | | DR 9 ID = 1.023 | | DR 11 ID = 1.077 | |
| Flow *gpm* | Head Loss *ft/100 ft* | Velocity *ft/s* | Head Loss *ft/100 ft* | Velocity *ft/s* | Head Loss *ft/100 ft* | Velocity *ft/s* | Head Loss *ft/100 ft* | Velocity *ft/s* |
| 2 | 0.88 | 1.07 | 0.68 | 0.96 | 0.41 | 0.78 | 0.32 | 0.70 |
| 3 | 1.86 | 1.60 | 1.45 | 1.44 | 0.87 | 1.17 | 0.68 | 1.06 |
| 4 | 3.17 | 2.13 | 2.47 | 1.93 | 1.48 | 1.56 | 1.15 | 1.41 |
| 5 | 4.79 | 2.67 | 3.73 | 2.41 | 2.24 | 1.95 | 1.74 | 1.76 |
| 6 | 6.71 | 3.20 | 5.23 | 2.89 | 3.14 | 2.34 | 2.44 | 2.11 |
| 7 | 8.93 | 3.73 | 6.96 | 3.37 | 4.17 | 2.73 | 3.25 | 2.47 |
| 8 | 11.44 | 4.27 | 8.91 | 3.85 | 5.34 | 3.12 | 4.16 | 2.82 |
| 9 | 14.22 | 4.80 | 11.08 | 4.33 | 6.64 | 3.51 | 5.17 | 3.17 |

*(continued)*

**Table B-26** Polyethylene pipe—C=130—1 in. nominal diameter (*Continued*)

| *C* = 130 | ID = Inside diameter* | | | | | | | |
|---|---|---|---|---|---|---|---|---|
| | Copper Tubing Sizes | | | | Iron Pipe Sizes | | | |
| | DR 9 ID = 0.875 | | DR 11 ID = 0.921 | | DR 9 ID = 1.023 | | DR 11 ID = 1.077 | |
| Flow *gpm* | Head Loss *ft/100 ft* | Velocity *ft/s* | Head Loss *ft/100 ft* | Velocity *ft/s* | Head Loss *ft/100 ft* | Velocity *ft/s* | Head Loss *ft/100 ft* | Velocity *ft/s* |
| 10 | 17.29 | 5.34 | 13.47 | 4.82 | 8.08 | 3.90 | 6.29 | 3.52 |
| 12 | 24.23 | 6.40 | 18.88 | 5.78 | 11.32 | 4.68 | 8.81 | 4.23 |
| 14 | 32.24 | 7.47 | 25.12 | 6.74 | 15.06 | 5.46 | 11.72 | 4.93 |
| 16 | 41.28 | 8.54 | 32.16 | 7.71 | 19.28 | 6.25 | 15.01 | 5.63 |
| 18 | 51.34 | 9.60 | 40.00 | 8.67 | 23.98 | 7.03 | 18.67 | 6.34 |
| 20 | | | 48.62 | 9.63 | 29.15 | 7.81 | 22.69 | 7.04 |
| 25 | | | | | 44.07 | 9.76 | 34.30 | 8.80 |
| 30 | | | | | | | 48.08 | 10.57 |

*ANSI/AWWA C901 (Average outside diameter less minimum wall thickness)

**Table B-27** Polyethylene pipe—C=130—1¼ in. nominal diameter

| *C* = 130 | ID = Inside diameter* | | | | | | | |
|---|---|---|---|---|---|---|---|---|
| | Copper Tubing Sizes | | | | Iron Pipe Sizes | | | |
| | DR 9 ID = 1.069 | | DR 11 ID = 1.125 | | DR 9 ID = 1.292 | | DR 11 ID = 1.358 | |
| Flow *gpm* | Head Loss *ft/100 ft* | Velocity *ft/s* | Head Loss *ft/100 ft* | Velocity *ft/s* | Head Loss *ft/100 ft* | Velocity *ft/s* | Head Loss *ft/100 ft* | Velocity *ft/s* |
| 5 | 1.81 | 1.79 | 1.41 | 1.61 | 0.72 | 1.22 | 0.56 | 1.11 |
| 6 | 2.53 | 2.14 | 1.97 | 1.94 | 1.01 | 1.47 | 0.79 | 1.33 |
| 7 | 3.37 | 2.50 | 2.63 | 2.26 | 1.34 | 1.71 | 1.05 | 1.55 |
| 8 | 4.31 | 2.86 | 3.36 | 2.58 | 1.71 | 1.96 | 1.34 | 1.77 |
| 9 | 5.36 | 3.22 | 4.18 | 2.90 | 2.13 | 2.20 | 1.67 | 1.99 |
| 10 | 6.52 | 3.57 | 5.08 | 3.23 | 2.59 | 2.45 | 2.03 | 2.22 |
| 12 | 9.14 | 4.29 | 7.13 | 3.87 | 3.63 | 2.94 | 2.85 | 2.66 |
| 15 | 13.81 | 5.36 | 10.77 | 4.84 | 5.49 | 3.67 | 4.31 | 3.32 |
| 20 | 23.53 | 7.15 | 18.35 | 6.46 | 9.35 | 4.89 | 7.34 | 4.43 |
| 25 | 35.57 | 8.94 | 27.74 | 8.07 | 14.14 | 6.12 | 11.09 | 5.54 |
| 30 | 49.86 | 10.72 | 38.88 | 9.68 | 19.81 | 7.34 | 15.54 | 6.65 |
| 35 | | | 51.72 | 11.30 | 26.36 | 8.57 | 20.68 | 7.75 |
| 40 | | | | | 33.75 | 9.79 | 26.48 | 8.86 |
| 45 | | | | | 41.98 | 11.01 | 32.94 | 9.97 |
| 50 | | | | | | | 40.03 | 11.08 |

*ANSI/AWWA C901 (Average outside diameter less minimum wall thickness)

**Table B-28** Polyethylene pipe—C=130—1½ in. nominal diameter

| *C* = 130 | ID = Inside diameter* | | | | | | | |
|---|---|---|---|---|---|---|---|---|
| | Copper Tubing Sizes | | | | Iron Pipe Sizes | | | |
| | DR 9 ID = 1.263 | | DR 11 ID = 1.329 | | DR 9 ID = 1.478 | | DR 11 ID = 1.554 | |
| Flow *gpm* | Head Loss *ft/100 ft* | Velocity *ft/s* | Head Loss *ft/100 ft* | Velocity *ft/s* | Head Loss *ft/100 ft* | Velocity *ft/s* | Head Loss *ft/100 ft* | Velocity *ft/s* |
| 8 | 1.91 | 2.05 | 1.49 | 1.85 | 0.89 | 1.50 | 0.70 | 1.35 |
| 9 | 2.38 | 2.30 | 1.86 | 2.08 | 1.11 | 1.68 | 0.87 | 1.52 |
| 10 | 2.89 | 2.56 | 2.26 | 2.31 | 1.35 | 1.87 | 1.05 | 1.69 |
| 12 | 4.06 | 3.07 | 3.16 | 2.78 | 1.89 | 2.24 | 1.48 | 2.03 |
| 15 | 6.13 | 3.84 | 4.78 | 3.47 | 2.85 | 2.80 | 2.23 | 2.54 |
| 20 | 10.44 | 5.12 | 8.15 | 4.63 | 4.86 | 3.74 | 3.80 | 3.38 |
| 25 | 15.79 | 6.40 | 12.32 | 5.78 | 7.34 | 4.67 | 5.75 | 4.23 |
| 30 | 22.13 | 7.68 | 17.27 | 6.94 | 10.29 | 5.61 | 8.06 | 5.07 |
| 35 | 29.44 | 8.96 | 22.97 | 8.09 | 13.69 | 6.54 | 10.72 | 5.92 |
| 40 | 37.70 | 10.24 | 29.42 | 9.25 | 17.53 | 7.48 | 13.73 | 6.77 |
| 45 | | | 36.59 | 10.41 | 21.81 | 8.41 | 17.08 | 7.61 |
| 50 | | | | | 26.50 | 9.35 | 20.76 | 8.46 |
| 55 | | | | | 31.62 | 10.28 | 24.77 | 9.30 |
| 60 | | | | | | | 29.10 | 10.15 |

*ANSI/AWWA C901 (Average outside diameter less minimum wall thickness)

**Table B-29** Polyethylene pipe—C=130—2 in. nominal diameter

| *C* = 130 | ID = Inside diameter* | | | | | | | |
|---|---|---|---|---|---|---|---|---|
| | Copper Tubing Sizes | | | | Iron Pipe Sizes | | | |
| | DR 9 ID = 1.653 | | DR 11 ID = 1.739 | | DR 9 ID = 1.847 | | DR 11 ID = 1.943 | |
| Flow *gpm* | Head Loss *ft/100 ft* | Velocity *ft/s* | Head Loss *ft/100 ft* | Velocity *ft/s* | Head Loss *ft/100 ft* | Velocity *ft/s* | Head Loss *ft/100 ft* | Velocity *ft/s* |
| 10 | 0.78 | 1.50 | 0.61 | 1.35 | 0.45 | 1.20 | 0.36 | 1.08 |
| 12 | 1.09 | 1.79 | 0.85 | 1.62 | 0.64 | 1.44 | 0.50 | 1.30 |
| 14 | 1.45 | 2.09 | 1.14 | 1.89 | 0.85 | 1.68 | 0.66 | 1.51 |
| 16 | 1.86 | 2.39 | 1.46 | 2.16 | 1.09 | 1.92 | 0.85 | 1.73 |
| 18 | 2.32 | 2.69 | 1.81 | 2.43 | 1.35 | 2.16 | 1.05 | 1.95 |
| 20 | 2.82 | 2.99 | 2.20 | 2.70 | 1.64 | 2.39 | 1.28 | 2.16 |
| 25 | 4.26 | 3.74 | 3.33 | 3.38 | 2.48 | 2.99 | 1.94 | 2.71 |
| 30 | 5.97 | 4.49 | 4.66 | 4.05 | 3.48 | 3.59 | 2.72 | 3.25 |
| 35 | 7.94 | 5.23 | 6.20 | 4.73 | 4.62 | 4.19 | 3.61 | 3.79 |
| 40 | 10.17 | 5.98 | 7.94 | 5.40 | 5.92 | 4.79 | 4.63 | 4.33 |
| 45 | 12.64 | 6.73 | 9.88 | 6.08 | 7.36 | 5.39 | 5.75 | 4.87 |
| 50 | 15.37 | 7.48 | 12.00 | 6.75 | 8.95 | 5.99 | 6.99 | 5.41 |
| 55 | 18.33 | 8.22 | 14.32 | 7.43 | 10.68 | 6.59 | 8.34 | 5.95 |

*(continued)*

**Table B-29** Polyethylene pipe—C=130—2 in. nominal diameter (*Continued*)

| *C* = 130 | ID = Inside diameter* | | | | | | | |
|---|---|---|---|---|---|---|---|---|
| | Copper Tubing Sizes | | | | Iron Pipe Sizes | | | |
| | DR 9 ID = 1.653 | | DR 11 ID = 1.739 | | DR 9 ID = 1.847 | | DR 11 ID = 1.943 | |
| Flow *gpm* | Head Loss *ft/100 ft* | Velocity *ft/s* | Head Loss *ft/100 ft* | Velocity *ft/s* | Head Loss *ft/100 ft* | Velocity *ft/s* | Head Loss *ft/100 ft* | Velocity *ft/s* |
| 60 | 21.54 | 8.97 | 16.83 | 8.10 | 12.55 | 7.18 | 9.80 | 6.49 |
| 70 | 28.66 | 10.47 | 22.38 | 9.46 | 16.69 | 8.38 | 13.04 | 7.57 |
| 80 | | | 28.66 | 10.81 | 21.37 | 9.58 | 16.70 | 8.66 |
| 90 | | | | | 26.58 | 10.78 | 20.77 | 9.74 |
| 100 | | | | | | | 25.24 | 10.82 |

*ANSI/AWWA C901 (Average outside diameter less minimum wall thickness)

**Table B-30** Polyethylene pipe—C=130—3 in. nominal diameter

| *C* = 130 | ID = Inside diameter* | | | | | | | |
|---|---|---|---|---|---|---|---|---|
| | Copper Tubing Sizes | | | | Iron Pipe Sizes | | | |
| | DR 9 ID = N/A† | | DR 11 ID = N/A | | DR 9 ID = 2.722 | | DR 11 ID = 2.864 | |
| Flow *gpm* | Head Loss *ft/100 ft* | Velocity *ft/s* | Head Loss *ft/100 ft* | Velocity *ft/s* | Head Loss *ft/100 ft* | Velocity *ft/s* | Head Loss *ft/100 ft* | Velocity *ft/s* |
| 20 | N/A | N/A | N/A | N/A | 0.25 | 1.10 | 0.19 | 1.00 |
| 30 | N/A | N/A | N/A | N/A | 0.53 | 1.65 | 0.41 | 1.49 |
| 40 | N/A | N/A | N/A | N/A | 0.90 | 2.21 | 0.70 | 1.99 |
| 50 | N/A | N/A | N/A | N/A | 1.35 | 2.76 | 1.06 | 2.49 |
| 60 | N/A | N/A | N/A | N/A | 1.90 | 3.31 | 1.48 | 2.99 |
| 70 | N/A | N/A | N/A | N/A | 2.52 | 3.86 | 1.97 | 3.49 |
| 80 | N/A | N/A | N/A | N/A | 3.23 | 4.41 | 2.52 | 3.98 |
| 90 | N/A | N/A | N/A | N/A | 4.02 | 4.96 | 3.14 | 4.48 |
| 100 | N/A | N/A | N/A | N/A | 4.89 | 5.51 | 3.82 | 4.98 |
| 110 | N/A | N/A | N/A | N/A | 5.83 | 6.06 | 4.55 | 5.48 |
| 120 | N/A | N/A | N/A | N/A | 6.85 | 6.62 | 5.35 | 5.98 |
| 130 | N/A | N/A | N/A | N/A | 7.94 | 7.17 | 6.20 | 6.47 |
| 140 | N/A | N/A | N/A | N/A | 9.11 | 7.72 | 7.11 | 6.97 |
| 150 | N/A | N/A | N/A | N/A | 10.36 | 8.27 | 8.08 | 7.47 |
| 160 | | | N/A | N/A | 11.67 | 8.82 | 9.11 | 7.97 |
| 170 | | | N/A | N/A | 13.06 | 9.37 | 10.19 | 8.47 |
| 180 | | | | | 14.51 | 9.92 | 11.33 | 8.96 |
| 190 | | | | | 16.04 | 10.48 | 12.52 | 9.46 |
| 200 | | | | | | | 13.77 | 9.96 |
| 210 | | | | | | | 15.07 | 10.46 |

*ANSI/AWWA C901 (Average outside diameter less minimum wall thickness)
†NA = not applicable

**Table B-31** Cross-linked polyethylene (PEX) pipe—C=130—½ in. nominal diameter, 0.625

| *C* = 130 | ID = Inside diameter* | | |
|---|---|---|---|
| | Copper Tubing Sizes | | |
| | DR 9<br>ID = 0.485 | | Thickness = 0.07 |
| Flow<br>*gpm* | Head Loss<br>*ft/100 ft* | Velocity<br>*ft/s* | |
| 1 | 4.31 | 1.74 | |
| 2 | 15.54 | 3.47 | |
| 3 | 32.93 | 5.21 | |
| 4 | 56.09 | 6.95 | |
| 5 | 84.80 | 8.68 | |
| 6 | 118.85 | 10.42 | |

*ANSI/AWWA C904-06 (Average outside diameter less minimum wall thickness)

**Table B-32** Cross-linked polyethylene (PEX) pipe—C=130—⅝ in. nominal diameter, 0.75

| *C* = 130 | ID = Inside diameter* | | |
|---|---|---|---|
| | Copper Tubing Sizes | | |
| | DR 9<br>ID = 0.583333 | | Thickness = 0.083333 |
| Flow<br>*gpm* | Head Loss<br>*ft/100 ft* | Velocity<br>*ft/s* | |
| 1 | 1.75 | 1.20 | |
| 2 | 6.32 | 2.40 | |
| 3 | 13.40 | 3.60 | |
| 4 | 22.83 | 4.80 | |
| 5 | 34.51 | 6.00 | |
| 6 | 48.36 | 7.20 | |
| 7 | 64.34 | 8.40 | |
| 8 | 82.39 | 9.60 | |
| 9 | 102.48 | 10.80 | |

*ANSI/AWWA C904-06 (Average outside diameter less minimum wall thickness)

**Table B-33** Cross-linked polyethylene (PEX) pipe—C=130—¾ in. nominal diameter, 0.875

| *C* = 130 | ID = Inside diameter* | | |
|---|---|---|---|
| | Copper Tubing Sizes | | |
| | DR 9<br>ID = 0.680556 | | Thickness = 0.097222 |
| Flow<br>*gpm* | Head Loss<br>*ft/100 ft* | Velocity<br>*ft/s* | |
| 1 | 0.83 | 0.88 | |
| 2 | 2.98 | 1.76 | |
| 3 | 6.32 | 2.65 | |
| 4 | 10.77 | 3.53 | |

*(continued)*

**Table B-33** Cross-linked polyethylene (PEX) pipe—C=130—¾ in. nominal diameter, 0.875 (*Continued*)

| *C* = 130 | ID = Inside diameter* | | |
|---|---|---|---|
| | Copper Tubing Sizes | | |
| | DR 9<br>ID = 0.680556 | | Thickness = 0.097222 |
| **Flow**<br>*gpm* | **Head Loss**<br>*ft/100 ft* | **Velocity**<br>*ft/s* | |
| 5 | 16.29 | 4.41 | |
| 6 | 22.83 | 5.29 | |
| 7 | 30.37 | 6.17 | |
| 8 | 38.89 | 7.06 | |
| 9 | 48.37 | 7.94 | |
| 10 | 58.79 | 8.82 | |
| 11 | 70.14 | 9.70 | |
| 12 | 82.40 | 10.58 | |

*ANSI/AWWA C904-06 (Average outside diameter less minimum wall thickness)

**Table B-34** Cross-linked polyethylene (PEX) pipe—C=130—1 in. nominal diameter, 1.125

| *C* = 130 | ID = Inside diameter* | | |
|---|---|---|---|
| | Copper Tubing Sizes | | |
| | DR 9<br>ID = 0.875 | | Thickness = 0.125 |
| **Flow**<br>*gpm* | **Head Loss**<br>*ft/100 ft* | **Velocity**<br>*ft/s* | |
| 1 | 0.24 | 0.53 | |
| 2 | 0.88 | 1.07 | |
| 3 | 1.86 | 1.60 | |
| 4 | 3.17 | 2.13 | |
| 5 | 4.79 | 2.67 | |
| 6 | 6.71 | 3.20 | |
| 7 | 8.93 | 3.73 | |
| 8 | 11.44 | 4.27 | |
| 9 | 14.22 | 4.80 | |
| 10 | 17.29 | 5.34 | |
| 11 | 20.62 | 5.87 | |
| 12 | 24.23 | 6.40 | |
| 13 | 28.10 | 6.94 | |
| 14 | 32.24 | 7.47 | |
| 15 | 36.63 | 8.00 | |
| 16 | 41.28 | 8.54 | |
| 17 | 46.18 | 9.07 | |
| 18 | 51.34 | 9.60 | |
| 19 | 56.75 | 10.14 | |

*ANSI/AWWA C904-06 (Average outside diameter less minimum wall thickness)

**Table B-35** Cross-linked polyethylene (PEX) pipe—C=130—1¼ in. nominal diameter, 1.375

| *C* = 130 | ID = Inside diameter* | | |
|---|---|---|---|
| | Copper Tubing Sizes | | |
| | DR 9<br>ID = 1.069444 | | Thickness = 0.152778 |
| **Flow**<br>*gpm* | **Head Loss**<br>*ft/100 ft* | **Velocity**<br>*ft/s* | |
| 1 | 0.09 | 0.36 | |
| 2 | 0.33 | 0.71 | |
| 3 | 0.70 | 1.07 | |
| 4 | 1.19 | 1.43 | |
| 5 | 1.80 | 1.79 | |
| 6 | 2.53 | 2.14 | |
| 7 | 3.36 | 2.50 | |
| 8 | 4.30 | 2.86 | |
| 9 | 5.35 | 3.21 | |
| 10 | 6.51 | 3.57 | |
| 11 | 7.76 | 3.93 | |
| 12 | 9.12 | 4.29 | |
| 13 | 10.58 | 4.64 | |
| 14 | 12.13 | 5.00 | |
| 15 | 13.78 | 5.36 | |
| 20 | 23.48 | 7.14 | |
| 25 | 35.50 | 8.93 | |
| 30 | 49.75 | 10.72 | |

*ANSI/AWWA C904-06 (Average outside diameter less minimum wall thickness)

**Table B-36** Cross-linked polyethylene (PEX) pipe—C=130—1½ in. nominal diameter, 1.625

| *C* = 130 | ID = Inside diameter* | | |
|---|---|---|---|
| | Copper Tubing Sizes | | |
| | DR 9<br>ID = 1.263889 | | Thickness = 0.180556 |
| **Flow**<br>*gpm* | **Head Loss**<br>*ft/100 ft* | **Velocity**<br>*ft/s* | |
| 1 | 0.04 | 0.26 | |
| 2 | 0.15 | 0.51 | |
| 3 | 0.31 | 0.77 | |
| 4 | 0.53 | 1.02 | |
| 5 | 0.80 | 1.28 | |
| 6 | 1.12 | 1.53 | |
| 7 | 1.49 | 1.79 | |
| 8 | 1.91 | 2.05 | |
| 9 | 2.37 | 2.30 | |
| 10 | 2.88 | 2.56 | |

*(continued)*

**Table B-36** Cross-linked polyethylene (PEX) pipe—C=130—1½ in. nominal diameter, 1.625 (*Continued*)

| *C* = 130 | ID = Inside diameter* | | |
|---|---|---|---|
| | Copper Tubing Sizes | | |
| | DR 9<br>ID = 1.263889 | | Thickness = 0.180556 |
| **Flow**<br>*gpm* | **Head Loss**<br>*ft/100 ft* | **Velocity**<br>*ft/s* | |
| 11 | 3.44 | 2.81 | |
| 12 | 4.04 | 3.07 | |
| 13 | 4.69 | 3.32 | |
| 14 | 5.38 | 3.58 | |
| 15 | 6.11 | 3.84 | |
| 20 | 10.41 | 5.11 | |
| 25 | 15.73 | 6.39 | |
| 30 | 22.05 | 7.67 | |
| 35 | 29.34 | 8.95 | |
| 40 | 37.57 | 10.23 | |

*ANSI/AWWA C904-06 (Average outside diameter less minimum wall thickness)

**Table B-37** Cross-linked polyethylene (PEX) pipe—C=130—2 in. nominal diameter, 2.125

| *C* = 130 | ID = Inside diameter* | | |
|---|---|---|---|
| | Copper Tubing Sizes | | |
| | DR 9<br>ID = 1.652778 | | Thickness = 0.236111 |
| **Flow**<br>*gpm* | **Head Loss**<br>*ft/100 ft* | **Velocity**<br>*ft/s* | |
| 1 | 0.01 | 0.15 | |
| 2 | 0.04 | 0.30 | |
| 3 | 0.08 | 0.45 | |
| 4 | 0.14 | 0.60 | |
| 5 | 0.22 | 0.75 | |
| 10 | 0.78 | 1.50 | |
| 15 | 1.65 | 2.24 | |
| 20 | 2.82 | 2.99 | |
| 25 | 4.26 | 3.74 | |
| 30 | 5.97 | 4.49 | |
| 35 | 7.94 | 5.23 | |
| 40 | 10.17 | 5.98 | |
| 45 | 12.65 | 6.73 | |
| 50 | 15.38 | 7.48 | |
| 55 | 18.35 | 8.22 | |
| 60 | 21.55 | 8.97 | |
| 65 | 25.00 | 9.72 | |
| 70 | 28.67 | 10.47 | |

*ANSI/AWWA C904-06 (Average outside diameter less minimum wall thickness)

**Table B-38** Cross-linked polyethylene (PEX) pipe—C=130—2½ in. nominal diameter, 2.625

| *C* = 130 | ID = Inside diameter* | | |
|---|---|---|---|
| | Copper Tubing Sizes | | |
| | DR9<br>ID = 2.041667 | | Thickness = 0.291667 |
| **Flow**<br>*gpm* | **Head Loss**<br>*ft/100 ft* | **Velocity**<br>*ft/s* | |
| 10 | 0.28 | 0.98 | |
| 15 | 0.59 | 1.47 | |
| 20 | 1.01 | 1.96 | |
| 25 | 1.52 | 2.45 | |
| 30 | 2.13 | 2.94 | |
| 35 | 2.84 | 3.43 | |
| 40 | 3.63 | 3.92 | |
| 45 | 4.52 | 4.41 | |
| 50 | 5.49 | 4.90 | |
| 55 | 6.56 | 5.39 | |
| 60 | 7.70 | 5.88 | |
| 65 | 8.93 | 6.37 | |
| 70 | 10.25 | 6.86 | |
| 75 | 11.64 | 7.35 | |
| 80 | 13.12 | 7.84 | |
| 85 | 14.68 | 8.33 | |
| 90 | 16.32 | 8.82 | |
| 95 | 18.04 | 9.31 | |
| 100 | 19.83 | 9.80 | |
| 105 | 21.71 | 10.29 | |

*ANSI/AWWA C904-06 (Average outside diameter less minimum wall thickness)

**Table B-39** Cross-linked polyethylene (PEX) pipe—C=130—3 in. nominal diameter, 3.125

| *C* = 130 | ID = Inside diameter* | | |
|---|---|---|---|
| | Copper Tubing Sizes | | |
| | DR 9<br>ID = 2.430556 | | Thickness = 0.347222 |
| **Flow**<br>*gpm* | **Head Loss**<br>*ft/100 ft* | **Velocity**<br>*ft/s* | |
| 50 | 2.35 | 3.46 | |
| 55 | 2.80 | 3.80 | |
| 60 | 3.29 | 4.15 | |
| 65 | 3.82 | 4.49 | |
| 70 | 4.38 | 4.84 | |
| 75 | 4.98 | 5.19 | |
| 80 | 5.61 | 5.53 | |
| 85 | 6.28 | 5.88 | |

*(continued)*

**Table B-39** Cross-linked polyethylene (PEX) pipe—C=130—3 in. nominal diameter, 3.125 (*Continued*)

| *C* = 130 | **ID = Inside diameter*** | | |
|---|---|---|---|
| | **Copper Tubing Sizes** | | |
| | **DR 9<br>ID = 2.430556** | | **Thickness = 0.347222** |
| **Flow**<br>*gpm* | **Head Loss**<br>*ft/100 ft* | **Velocity**<br>*ft/s* | |
| **90** | 6.98 | 6.22 | |
| **95** | 7.72 | 6.57 | |
| **100** | 8.48 | 6.91 | |
| **105** | 9.29 | 7.26 | |
| **110** | 10.12 | 7.61 | |
| **115** | 10.99 | 7.95 | |
| **120** | 11.89 | 8.30 | |
| **125** | 12.83 | 8.64 | |
| **130** | 13.79 | 8.99 | |
| **135** | 14.79 | 9.33 | |
| **140** | 15.82 | 9.68 | |
| **145** | 16.88 | 10.03 | |

*ANSI/AWWA C904-06 (Average outside diameter less minimum wall thickness)

## REFERENCES

ASTM International (ASTM). B88. Standard Specification for Seamless Copper Water Tube. West Conshohocken, Pa.: ASTM.

ASTM. D2241. Standard Specification for Poly(Vinyl Chloride) (PVC) Pressure-Rated Pipe (SDR Series). West Conshohocken, Pa.: ASTM.

American National Standards Institute/American Water Works Association (ANSI/AWWA). 1997. C900. Polyvinyl Chloride (PVC) Pressure Pipe and Fabricated Fittings, 4 in. through 12 in. (100 mm through 300 mm), for Water Distribution. Denver: AWWA.

ANSI/AWWA. C901. Polyethylene (PE) Pressure Pipe and Tubing, ¾ in. (19 mm) through 3 in. (76 mm), for Water Service. Denver: AWWA.

ANSI/AWWA. 2006. C904. Crosslinked Polyethylene (PEX) Pressure Tubing, ½ in. (13 mm) through 3 in. (76 mm), for Water Service. Denver: AWWA.

# INDEX

Note: *f*. indicates figure

## M



## N



## O



## P



## R



## S



## T

## U



## V



## W

This page intentionally blank.

# AWWA Manuals

M65, *On-Site Generation of Hypochlorite,* #30065

M66, *Cylinder and Vane Actuators and Controls—Design and Installation,* #30066

M68, *Water Quality in Distribution Systems,* #30068

M69, *Inland Desalination and Concentrate Management,* #30069

M77, *Condition Assessment of Water Mains,* #30077

www.ingramcontent.com/pod-product-compliance
Ingram Content Group UK Ltd.
Pitfield, Milton Keynes, MK11 3LW, UK
UKHW052012041026
15590UKWH00008B/2297

9 781647 171247